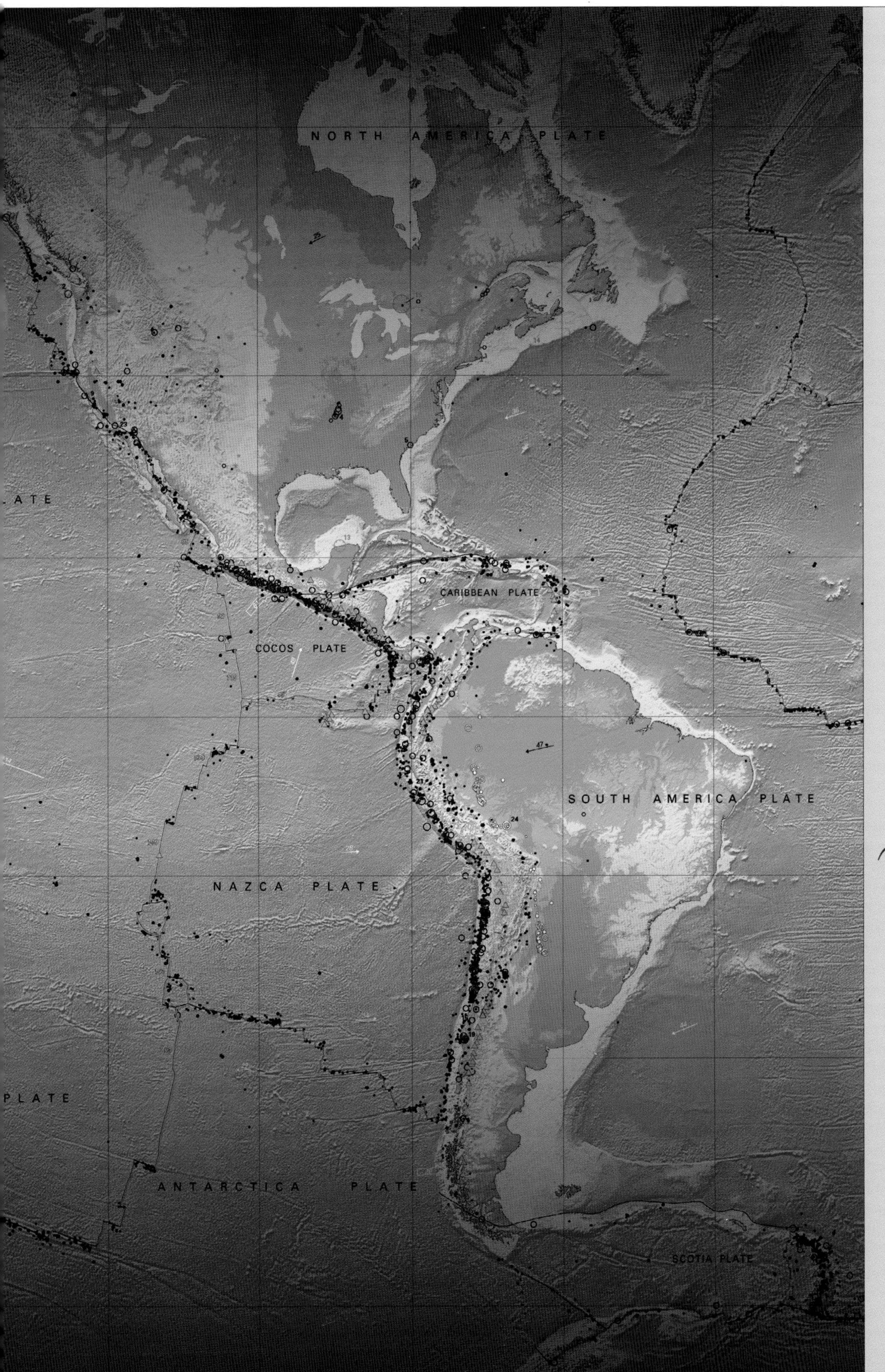

The collection of maps in this year's ESRI Map Book is one of the best ever created with geographic information system (GIS) technology. These maps demonstrate how our users are applying geographic information and analysis to assist in nearly every field of human endeavor. The use of GIS technology is introducing a new way of thinking and doing things. I like to refer to this as the geographic approach.

This approach involves using geography (the science of our world) as a framework to integrate all the important factors associated with problem solving—a kind of holistic approach. I strongly believe this approach will be increasingly important for managing and planning our world.

I want to thank the users who have allowed ESRI to publish these amazing contributions. The geographic approach is truly making a difference in the world.

Warm regards,

Jack Dangermond

June 2007

Table of Contents

Natural Resources — Forestry

Natural Resources — Mining and Earth Science

Natural Resources — Petroleum

Natural Resources — Water

Planning and Engineering

Sustainable Development and Human Affairs

Tourism

Transportation

Utilities — Electric and Gas

Utilities — Water and Wastewater

2006 Newark Comprehensive Economic Development Strategy Maps

Newwork, LLC
Newark, New Jersey, USA
By Michael Saltzman, Dong-Yun Shin, and Sunha Park

Contact
Dong-Yun Shin
dshin@newworking.com

Software
ArcView, Adobe Illustrator, Adobe InDesign

Hardware
Apple, PC

Printer
HP Designjet 800ps

Data Source(s)
ESRI; Newwork, LLC; U.S. Census

Introduction

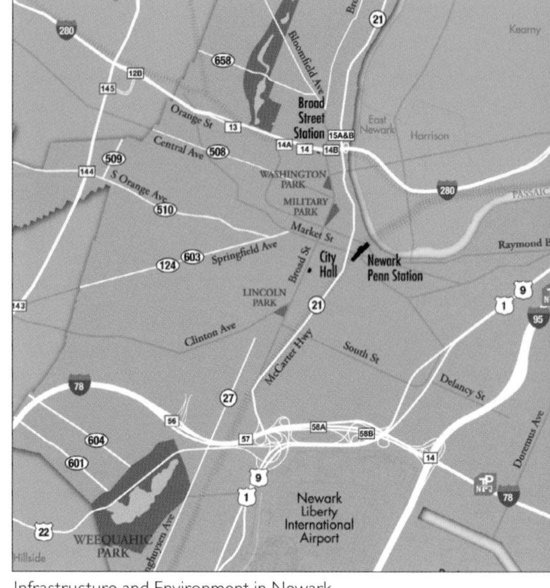

Infrastructure and Environment in Newark

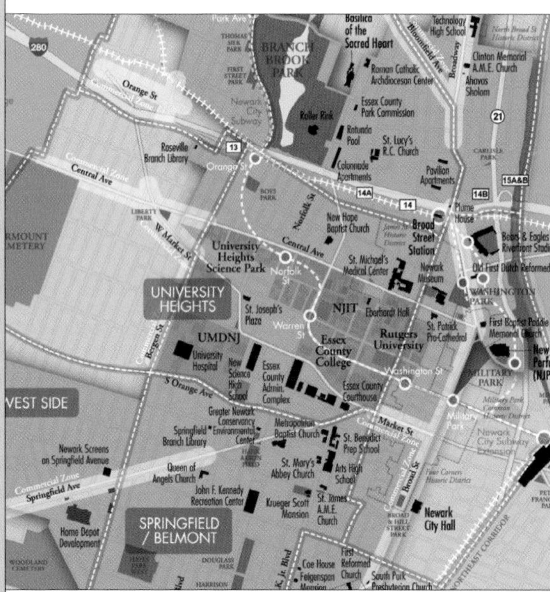

Newark Neighborhoods

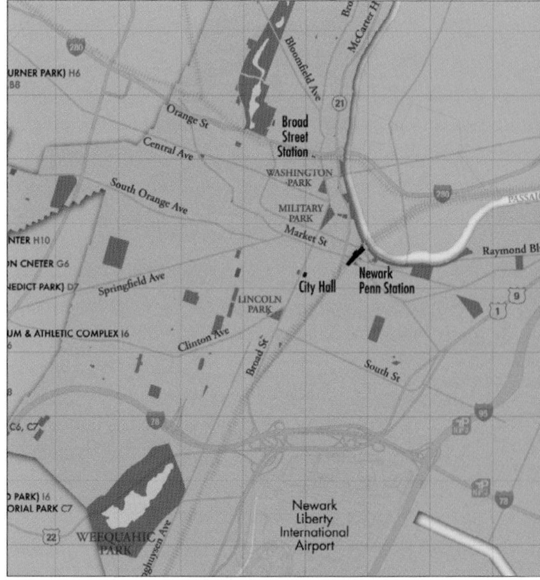

Cultural and Recreational Resources in Newark

The Newark Economy: Structure and Trends

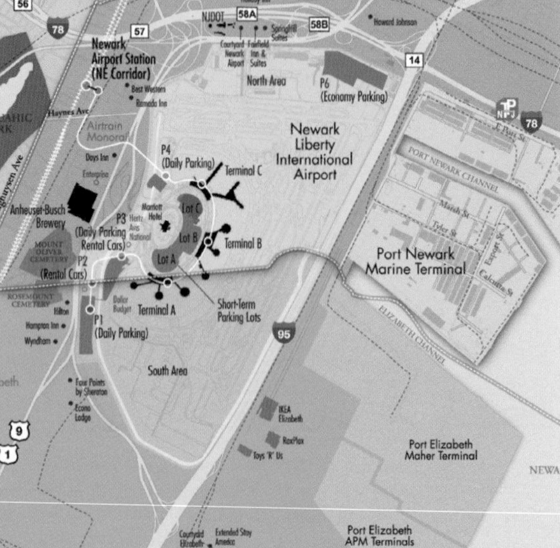

Land Use in Newark

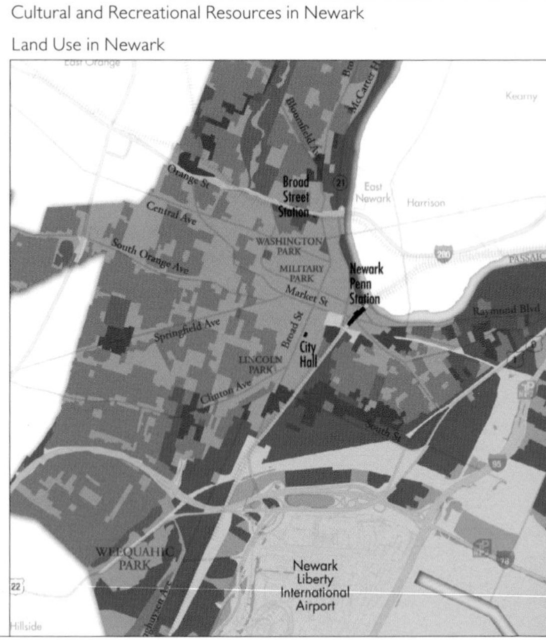

The Comprehensive Economic Development Strategy (CEDS) is a report by the City of Newark, New Jersey, to describe existing conditions in Newark and to formulate strategies and identify projects that will support the city's ongoing economic revitalization.

Twenty-five maps in the CEDS 2006 present Newark on a range of scales: from macro, showing the city as a transit hub in the Northeast; to micro, showing landmarks, historic districts, and commercial hubs in the city's five neighborhoods. Although each map delivers its own story, the 25 maps are linked to support the CEDS 2006.

The CEDS 2006 maps were developed by the planning and multimedia teams of Newwork, LLC, located in downtown Newark. The maps were enabled by the company's knowledge of Newark-based projects, its analysis and utilization of advanced GIS data, and its expertise in information design and graphic software.

Courtesy of Newwork, LLC.

City of Hampton
Hampton, Virginia, USA
By Jonah Adkins

Contact
Jonah Adkins
jadkins@hampton.gov

Software
ArcGIS Desktop 9.1

Printer
HP Designjet 1055cm

Data Source(s)
Hampton ArcSDE database

Interstate 64 & 664
Arterials
Collectors
Proposed
Residential
Hampton City Boundary
Trails & Canoe Routes
Tax Map Grid
Wetlands Features
Office Parks
Commercial Zoning
Parks & Recreation

The City of Hampton, Virginia, created its Street Centerline Map, using ArcGIS 9.1 software, to be the basic large-scale city map used by staff and citizens alike. It features mainly hand-placed, feature-linked annotation layers created in the ArcSDE database. A visually pleasing color palette was chosen to refine the look of the map. One of the more helpful features of the map is the very readable street index. The index corresponds to the municipal tax grid, which is the basis for 200 scale maps across the city. The index is alphabetized with major streets highlighted in red. Included on the map are the locations of office and industrial parks, commercially zoned areas, parks and recreation areas, police and fire departments, schools, wetlands, libraries, subdivisions, and government facilities such as NASA, Langley Air Force Base, and Fort Monroe Army Station. The inset of the downtown area shows the location of key city-government offices and public parking areas. This map also serves as a template for large-scale maps in the city as extra layers are added or subtracted as needed. The maps are distributed to city departments and sold to citizens upon request at minimal charge.

Courtesy of Jonah Adkins, City of Hampton, Virginia.

Mount St. Helens Reloaded

U.S. Geological Survey
Vancouver, Washington, USA
By David Ramsey, Steve Schilling, Joel Robinson, Julie Griswold, Rob Wardwell, and John Ewert

Contact
David Ramsey
dramsey@usgs.gov

Software
ArcGIS Desktop, ArcSDE, ArcScene

Hardware
PC and UNIX

Printer
HP Designjet 2500cp

Data Source(s)
USGS, NASA, ESRI

On May 18, 1980, Mount St. Helens, Washington, exploded in a spectacular and devastating eruption that shocked the world. The eruption, one of the most powerful in the history of the United States, lowered the mountain's summit elevation from 2,950 meters to 2,549 meters, leaving a north-facing, horseshoe-shaped crater more than two kilometers wide. Most of the forest surrounding the volcano was destroyed and 57 people were killed. After the 1980 eruption, Mount St. Helens remained active. A large lava dome episodically extruded in the center of the volcano's empty crater until 1986. During the two decades following the May 18, 1980, eruption, a new glacier, Crater Glacier, formed tongues of ice around the east and west sides of the lava dome in the deeply shaded niche between the lava dome and the south crater wall.

As the most active volcano in the Cascade Range with a complex 300,000-year history, Mount St. Helens erupted again in the fall of 2004 as a new period of dome building within the 1980 crater began. Lava erupted just south of the 1980–1986 lava dome, cracking and bulldozing Crater Glacier. The volume of the new lava dome, during less than two years of eruption, is roughly equal to that of the 1980–1986 lava dome, and it continues to grow.

GIS technology is an important part of the response to this ongoing eruption. GIS is being used to estimate extruded growth rates and volumes of the new lava dome; to catalog remote instruments used to monitor the volcano; to forecast the potential directions, extents, and ashfall amounts associated with volcanic ash plumes; and to analyze and visualize volcanic hazards associated with the eruption. These efforts are important in understanding the mechanics of the eruption, planning effective instrument deployment, and informing federal, state, and local officials of potential dangers posed by continuing volcanic activity.

ArcScene perspective view of shaded-relief image draped over high-resolution DEM

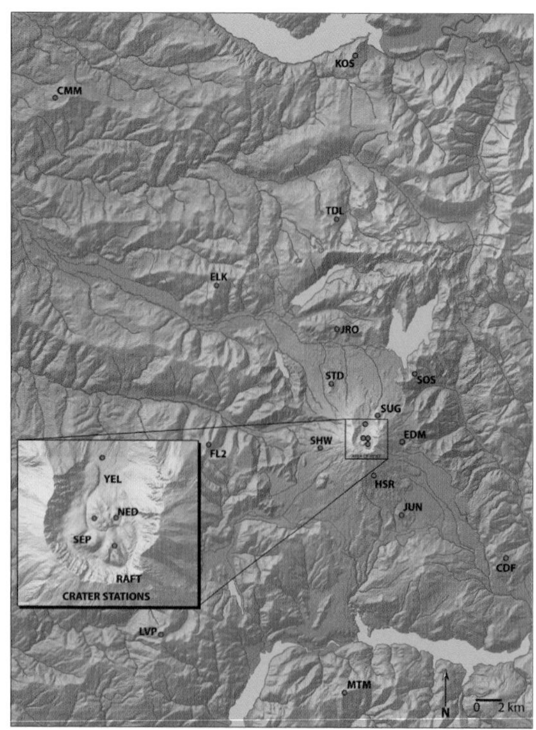

Mount St. Helens Seismic Station locations as of August 1, 2006

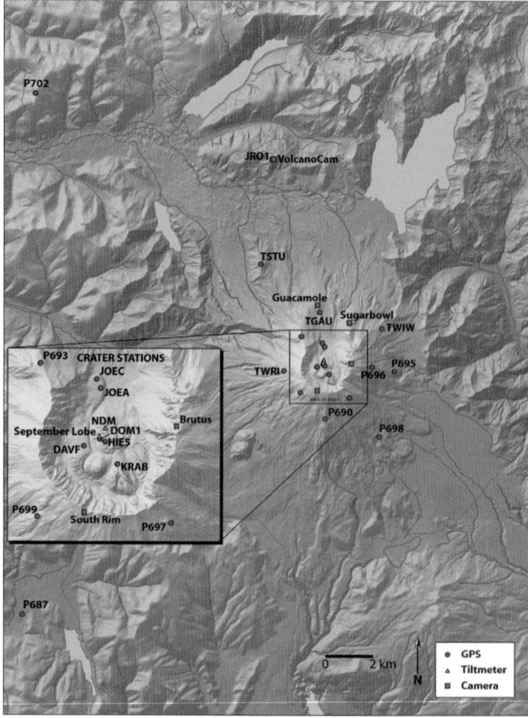

Mount St. Helens Deformation Station locations as of August 1, 2006

Boone County Planning Commission

Burlington, Kentucky, USA
By Louis R. Hill, Jr.

Contact

Louis R. Hill, Jr.
lrhill@boonecountyky.org

Software

ArcGIS Desktop 9.1, ArcGIS 3D Analyst, GPS Analyst, CorelDRAW 12

Hardware

Dell GX280

Printer

HP Designjet 3500cp

Data Source(s)

Boone County Planning Commission GIS Services

Arboretum Entrance Inset Map

The Boone County Arboretum is a 120-acre facility located in rural Boone County, Kentucky. Open to the public, the facility is unique in integrating an arboretum with active recreation (baseball diamonds, soccer fields) and passive recreation (walking/jogging trails, wildlife observation areas). The arboretum consists of more than 1,300 individually identified trees, more than 200 managed plant beds, eight featured points of interest, and an abundance of wooded and natural areas. Approximately 41,000 linear feet of underground irrigation lines and sprinklers help ensure that the arboretum's valuable investments in trees and plants will continue to live, even during drought-like conditions.

Visitors, even those with no horticultural expertise, can use the map for self-guided tours of the arboretum at any time of the year. The arboretum map uses an internal grid system to easily identify the various specimens on site. These identification systems are cross-referenced on both sides of the map and allow users to identify a specimen based on its ID number, grid location, or common name.

To better manage the constantly changing facility, the arboretum has implemented a combination of GIS and GPS technologies under the guidance of Boone County GIS. This allows the arboretum to continuously update its inventory of trees, shrubs, plantbeds, trails, woods, and wetlands.

Courtesy of Boone County Planning Commission.

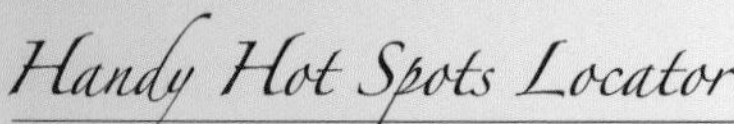

Handy Hot Spots Locator

Conservation International
Washington, D.C., USA
By Mark Denil

Contact
Mark Denil
m.denil@conservation.org

Software
ArcGIS Desktop 9.1

Printer
HP Designjet 5500ps

Data Source(s)
Conservation International, others

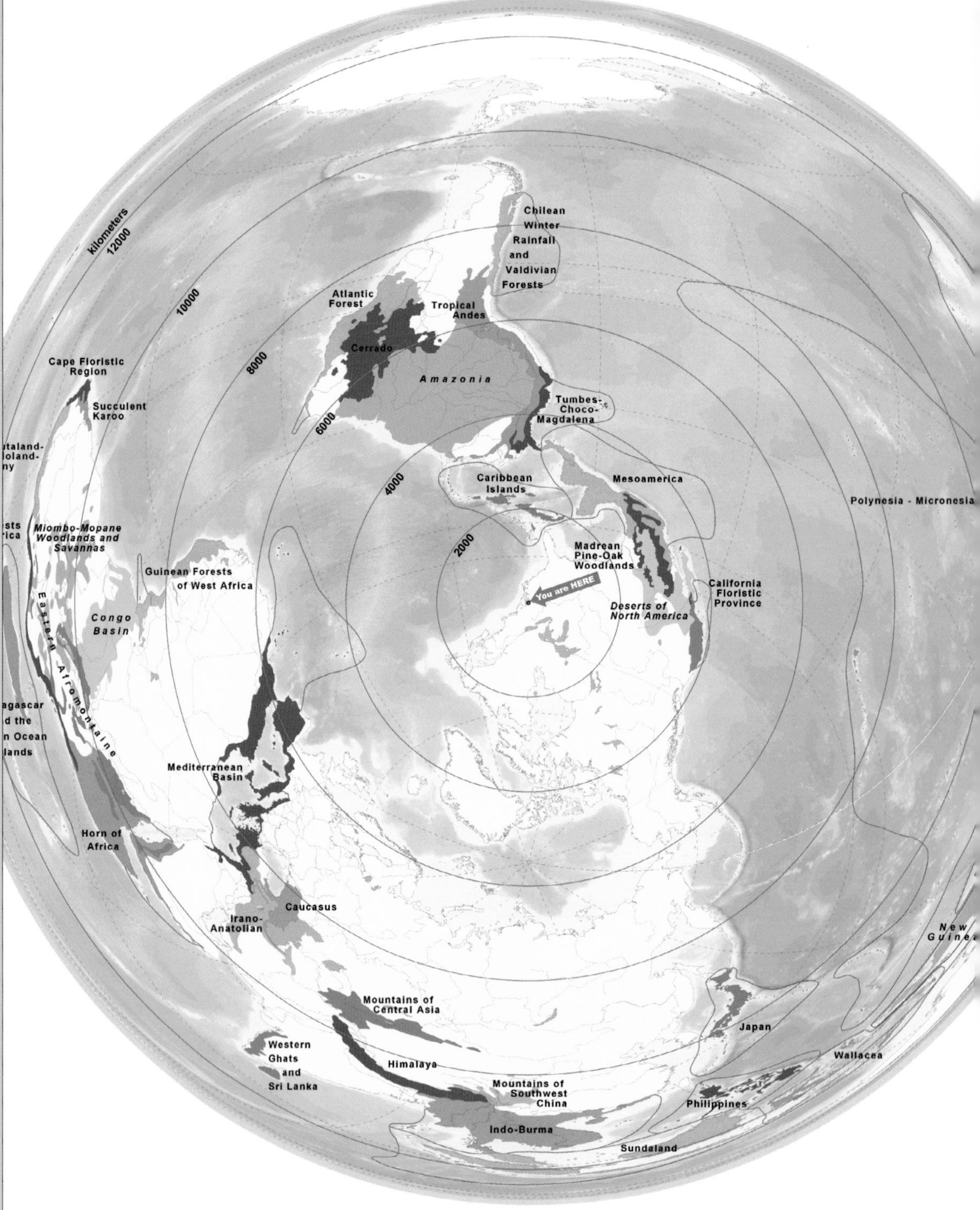

The hot-spots approach to the conservation of threatened ecosystems and species is a highly targeted strategy for tackling the overwhelming problem of biodiversity loss at the global level. On many world maps, the hot spots' locations and proximity may not seem obvious. This direction and proximity map is centered on the Conservation International head office.

A new analysis of the biodiversity hot spots identifies 34 regions worldwide where 75 percent of the planet's most threatened mammals, birds, and amphibians survive within habitat covering just 2.3 percent of the earth's surface. An estimated 50 percent of all vascular plants and 42 percent of terrestrial vertebrates exist only in these hot spots.

Courtesy of Conservation International.

Hydro Tower
Hydrants
Private Hydrants
Well Heads
Benchmarks
Storm Water Manholes
Storm Water Inlets
Storm Water Outlets
Storm Water Lead
Storm Water Pipe Network
Sanitary Manholes
Sanitary Pumping Stations
Sanitary Pipe Network
Water Valves
Water Pipes

Kitchener Infrastructure Atlas

Kitchener Emergency Response Atlas

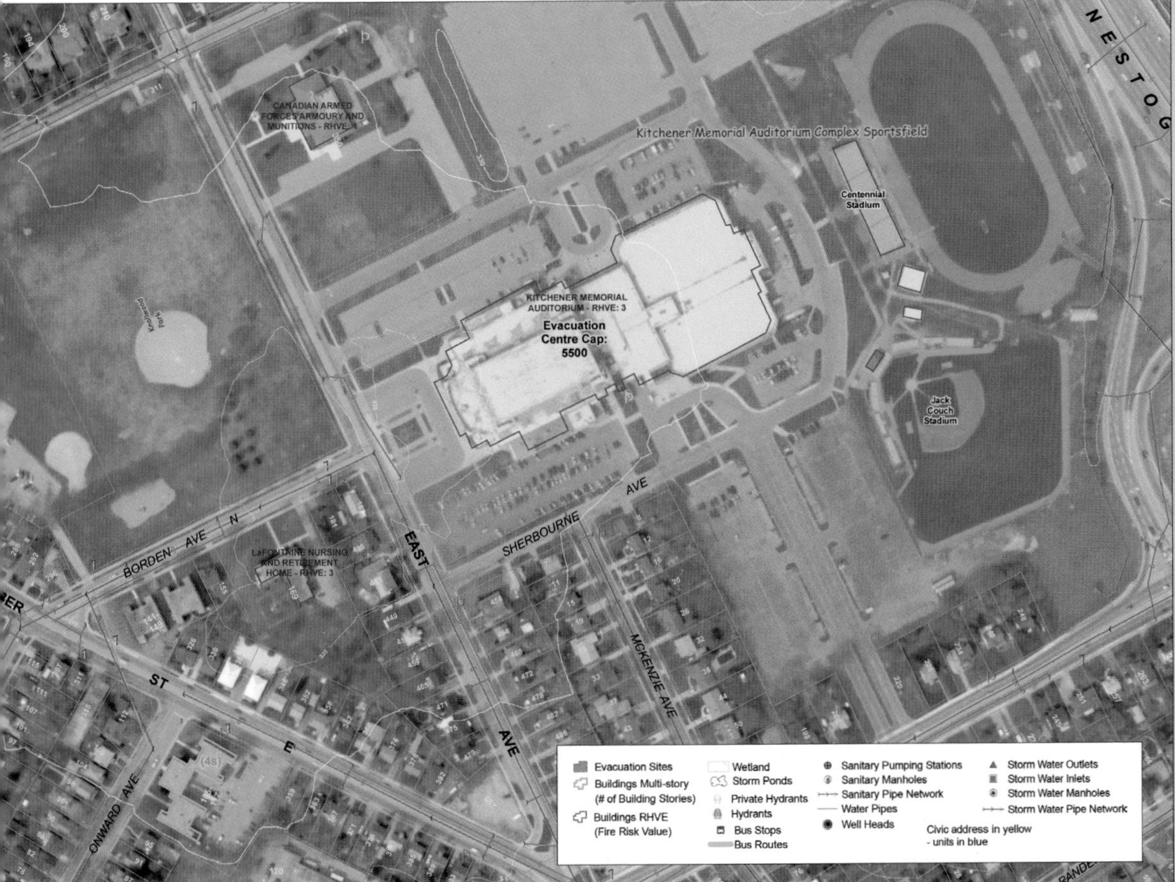

City of Kitchener

Kitchener, Ontario, Canada
By Melanie Wawryk

Contact

Jeff Ham
jeff.ham@kitchener.ca

Software

ArcGIS Desktop, ArcSDE

Printer

Xerox DC3535 Colour Laser Printer

Data Source(s)

City of Kitchener ArcSDE 9.1 database

Kitchener is a vibrant, confident, and cosmopolitan community with a population of 210,000 located in the heart of southwestern Ontario, Canada. The Kitchener 2006 Atlas Standard Edition displays information derived from the city's Enterprise GIS (ArcSDE database) including cadastral lines, building footprints, road pavement, driveways, sidewalks, fire hydrants, surface water, and land-use classes. More than 93,000 addresses are labeled in the atlas, including 8,500 horizontal units and 1,900 addresses on lots yet to be developed. All labeling for the map series is done dynamically using the ESRI Maplex extension. As a result, no labels in the series had to be converted to annotation to facilitate the production of a cartographically pleasing product.

The Kitchener 2006 Atlas Standard Edition is used primarily by city staff, both in the office and in the field, as a general quick reference, and can be updated on demand in addition to the annual editions. The City of Kitchener makes copies of the atlas available for the public to purchase. The GIS team at Kitchener also has produced three similar series of atlases. These display specific spatial information designed to support its internal Emergency Operations Centre, its city-owned property registry, and its engineering and public works operations staff.

Courtesy of the Corporation of the City of Kitchener.

Ordnance Survey

Southampton, UK
By Ordnance Survey

Contact

Customer Services
customerservices@ordnancesurvey.co.uk

Software

ArcGIS and other GIS software

Data Source(s)

OS MasterMap Topography Layer, OS MasterMap Integrated Transport Network Layer, OS MasterMap Address Layer 2

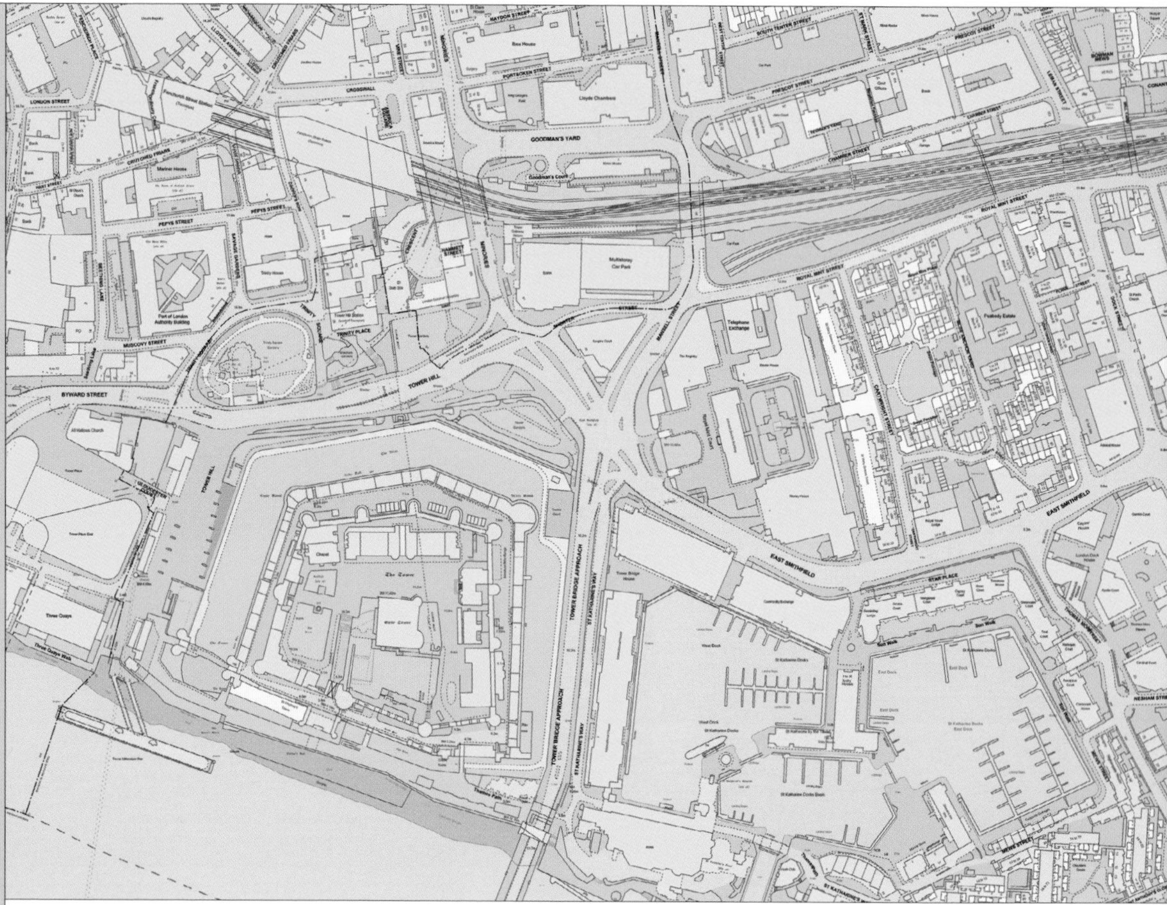

Tower of London

Buckingham Palace

OS MasterMap, provided by Ordnance Survey, is a digital representation of Great Britain containing more than 450 million uniquely identified geographic features. It is updated daily as a consistent framework for the referencing of geographic information in Great Britain. It is available in several data layers. First, the OS MasterMap Topography Layer is Ordnance Survey's flagship large-scale, polygonised, and seamless topographic database product. It is the most flexible and intelligent topographic information the Ordnance Survey has ever provided. Secondly, the OS MasterMap Integrated Transport Network (ITN) Layer provides the national definitive and intelligent multimodal transport network dataset for Great Britain. The third data layer is OS MasterMap Address Layer 2, which provides precise postal address coordinates across Great Britain and contains a range of new data, such as objects without postal addresses (OWPAs), multioccupancies without postal addresses (MOWPAs), alternative geographic/Welsh addresses, and building name aliases. OS MasterMap Address Layer is also available. Lastly, the OS MasterMap Imagery Layer is supplied to customers as ready-to-use, color-balanced, and fully edgematched orthorectified color aerial imagery data.

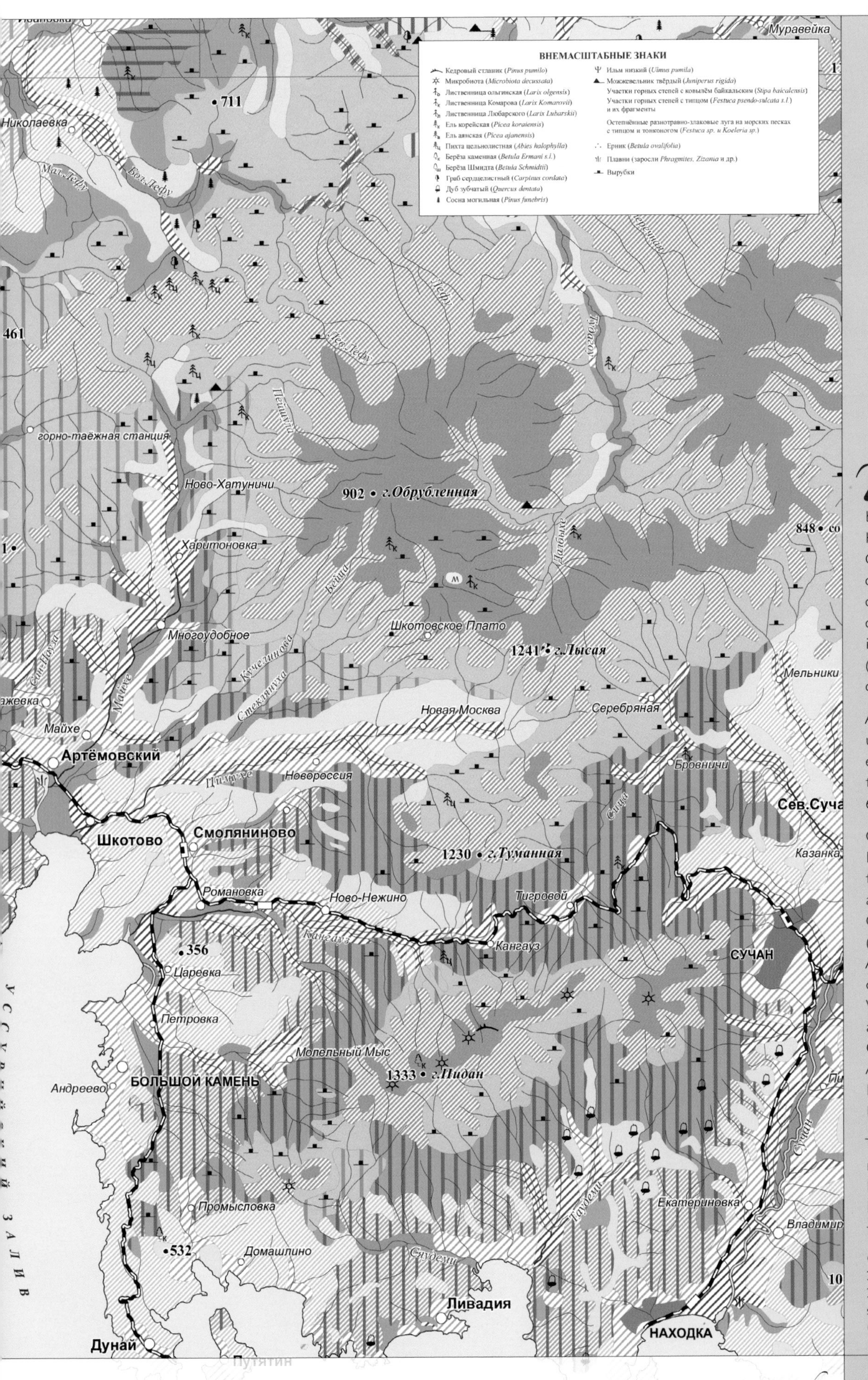

Pacific Institute of Geography
Far Eastern Branch of the Russian Academy of Sciences (FEB RAS)
Vladivostok, Russian Federation
By Sergey M. Krasnopeyev

Contact

Sergey M. Krasnopeyev
sergeikr@tig.dvo.ru

Software

ArcGIS Desktop 8.2

Printer

Indigo E-print 1000+

Data Source(s)

Forest management plans from 1989 to 2001

The Atlas of Primorskii Krai Forests is based on a geodatabase containing both cartographical and descriptive attributes of more than 470,000 forest stands in the "Primorskii Krai" (Russian Far East)—located at the juncture of Russia, China, and North Korea. Characteristics of state forest lands contained in forest management plans were entered into the database. The database was compiled by forest planning agencies for each national forest. It provided a mechanism for defining and mapping forest formations and grouped forest types (part of the hierarchical structure of the Sikhote-Alin forest classification system).

At the forest stand level, data for the last 10 years has been used to identify 160 to 170 forest types. For this reason, forest types can be grouped on the basis of the stand characteristics and the productivity of woody canopy, resulting in 63 grouped forest types.

Content of the atlas includes Vegetation of Primorskii Krai in 1955 (1:500,000); Habitat of the Grouped Forest Types of the Natural and Sustainable Secondary Forest Formations and Subformations (1:500,000); Distribution of the Area, Covered by a Forest, According to Predominant Species (1:500,000); Distribution of the Area, Covered by a Forest, According to Stand Aged Groups (1:500,000); Distribution of the Area, Covered by a Forest, According to the Timber Volume per Hectare (1:500,000); Specially Protected Areas and Forests with Protective Categories that Exclude the Commercial Timber Harvest (1:500,000); Volumes of the Above Ground Phytomass of the Forest Stands in the Main Forest Formations (1:1,500,000); and The History of the Forest Pests Outbreaks (1:1,500,000).

The atlas was prepared for publication by the collective of researchers from the Pacific Institute of Geography, the Institute of Biology and Soil Science of the Far Eastern Branch of the Russian Academy of Sciences (FEB RAS), and the Forest Service Agency.

 Edition was supported by the Russian Fund for Basic Research (Project #00-05-78053).

Collins Bartholomew, Ltd.

Glasgow, Scotland, UK
By Collins Bartholomew, Ltd.

Contact

John Allen
john.allen@harpercollins.co.uk

Software

ArcGIS Desktop 9.1

Printer

HP Designjet 5500

Data Source(s)

Collins Bartholomew World Premium dataset

*W*orld Scene is a georeferenced raster map covering the entire world. It is held in WGS 84 datum as a series of 48 seamless tiles, each of which has an image resolution of 300 dpi equating to approximately 425 meters on the ground at the equator. The map was created from Collins Bartholomew world vector data, which is held as ArcSDE feature classes on Oracle 9i and continuously maintained. These feature classes were initially symbolized using ArcMap and the Maplex Label Engine before converting labels to annotation to allow final manual tidying of text in busy areas.

The biggest challenge in creating this product was dealing with the variety of fonts and accented characters needed to satisfy the mapping specification. All place-names are shown in their local form. But because <FNT> (font) tags for those accents are not part of the Latin1 character set, they had to be manually applied to ensure a globally correct representation of names. Only a few, such as Vietnamese hooks, are not supported and are shown without diacritics. By setting up annotation feature classes and ArcMap templates, changes to the source vector data can quickly be reflected in the raster product to ensure that future versions are kept up to date.

Courtesy of Collins Bartholomew, Ltd.

Flat Planet Maps
Fallbrook, California, USA
By David Toney, GISP

Contact
David Toney, GISP
david@flatplanetmaps.com

Software
ArcGIS Desktop 8.3

Printer
HP Designjet 800

Data Source(s)
SANDAG, SanGIS

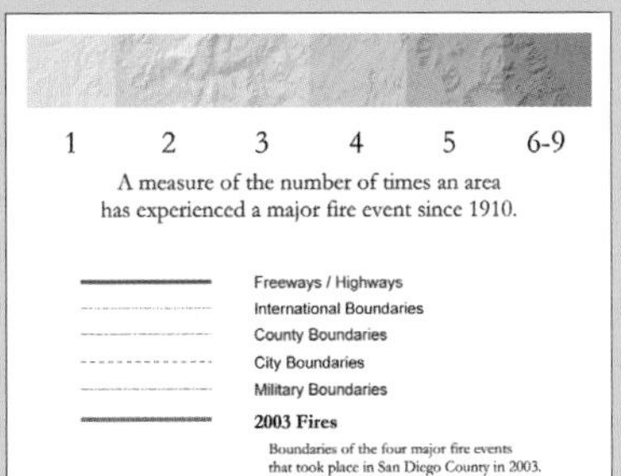

Wildland fire is a way of life in Southern California and can be devastating to businesses, residential communities, and wildlife. Application of frequency and interval models is an important component of regional wildland fire management. The main focus of this project was to investigate how often significant burn events have occurred in the past. The results of this study were implemented in the form of this map, which shows significant fires in San Diego County from 1910 to 2003.

Courtesy of David Toney, Flat Planet Maps.

This Dynamic Planet

Smithsonian Institution
Washington, D.C., USA
By Tom Simkin, Robert I. Tilling, Peter R. Vogt, Stephen H. Kirby, Paul Kimberly, and David Stewart

Contact
Daniel Cole
coled@si.edu

Software
ArcGIS Desktop

Printer
Printing press

Data Source(s)
Smithsonian Institution, USGS, U.S. Naval Research Laboratory

Scientists from the U.S. Geological Survey (USGS), Smithsonian Institution, and the U.S. Naval Research Laboratory published this new world map of volcanoes, earthquakes, impact craters, and plate tectonics. Its first two editions (1989, 1994) make it the best-selling map in the history of the USGS. The 1 x 1.5-meter map shows earth's most prominent features when viewed from a distance, and more detail upon closer inspection. The back of the map zooms in to show greater detail, with panels highlighting examples of earth's fundamental processes—plus time lines, references, and other resources to enhance the user's understanding. All of the map elements, plus interactive capability, are available at www.minerals.si.edu/tdpmap.

Courtesy of the USGS, Smithsonian Institution, and U.S. Naval Research Laboratory.

EURASIA PLATE

PHILIPPINE PLATE

PACIFIC PLATE

COCOS PL

NAZCA P

AUSTRALIA PLATE

PACIFIC PLATE

ANTARCTICA

ANTARCTICA PLATE

TOPOGRAPHY: DEPTH AND ELEVATION IN METERS

Sea-floor data from Scripps Institution of Oceanography Web site "Satellite Geodesy" (http://topex.ucsd.edu/marine_topo/mar_topo.html), accessed June 11, 1997; land-elevation data from U.S. Geological Survey (1997). Most lakes are shown in pale blue, with depth not implied. For the world's largest inland sea (Caspian) and deepest lake (Baikal), however, actual lake floor topography is shown using the same color scale as for the rest of the map.

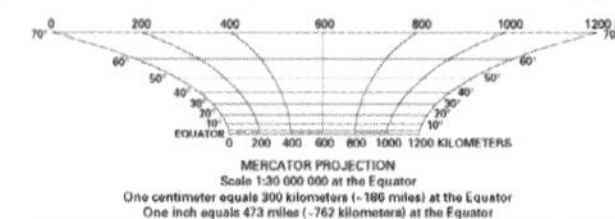

MERCATOR PROJECTION
Scale 1:30 000 000 at the Equator
One centimeter equals 300 kilometers (~186 miles) at the Equator
One inch equals 473 miles (~762 kilometers) at the Equator

Suggested citation: Simkin, Tom, Tilling, R.I., Vogt, P.R., Kirby, S.H., Kimberly, Paul, and Stewart, D.B., 2006, This dynamic planet: World map of volcanoes, earthquakes, impact craters, and plate tectonics: U.S. Geological Survey Geologic Investigations Series Map I-2800, 1 two-sided sheet, scale 1:30,000,000

Volcanoes—Data from Global Volcanism Program, Smithsonian Institution, Washington, D.C.; accessed at http://www.volcano.si.edu/world/summary.cfm , March 16, 2005

▲ Erupted A.D. 1900 through 2004
△ Erupted A.D. 1 through 1899
△ Erupted in Holocene time (past 10,000 years), but no known eruptions since A.D. 1
△ Uncertain Holocene activity and fumarolic activity

Impact Craters—Data from University of New Brunswick, Planetary and Space Science Centre, Earth Impact Database; accessed at http://www.unb.ca/passc/ImpactDatabase/, October 23, 2003 (also see Grieve, 1998). Geologic age span: 50 years to 2,400 million years. Crater diameter indicated below

- <10 km
- 10 to 70 km
- >70 km (shown at actual map scale)

Notable Events—Numbers next to a few symbols—of many thousands shown—denote especially noteworthy events, keyed to correspondingly numbered entries in tables found on the back of the map. These numbered events have produced devastating natural disasters, advanced scientific understanding, or piqued popular interest. They remind us that the map's small symbols may represent large and geologically significant events

19 Volcanoes
9 Earthquakes
23 Impact craters

Plate Tectonics

Divergent (sea-floor spreading) and transform fault boundaries—Red lines mark spreading centers where most of the world's volcanism takes place; thickness of lines indicates divergence rate, in four velocity ranges. White number is speed in millimeters per year (mm/yr) from DeMets and others (1994). The four spreading-rate ranges are <30 mm/yr; 30–59 mm/yr; 60–90 mm/yr; and >90 mm/yr. Thin black line marks the plate boundary, whether sea-floor spreading center or transform fault. On land, divergent boundaries are commonly diffuse zones (see interpretive map to the left); therefore, most are not shown. The only transform faults shown on land are those separating named plates

Plate motion—Data from Rice University Global Tectonics Group. Length of arrow is proportional to plate velocity, shown in millimeters per year. These approximate rates and directions are calculated from angular velocities with respect to hotspots, assumed to be relatively fixed in the mantle (see plate motion calculator at http://tectonics.rice.edu/hs3.html)

Plate convergence—More accurately known than "absolute" plate motion (above), convergence data are shown by arrows of uniform length showing direction and speed, in millimeters per year, relative to the plate across the boundary. Data from Charles DeMets (University of Wisconsin at Madison, written commun., 2003) and Bird (2003)

Earthquakes—Data from Engdahl and Villaseñor (2002). From 1900 through 1963, the data are complete for all earthquakes ≥6.5 magnitude; from 1964 through 1999, the data are complete for all earthquakes ≥5.0 magnitude. Most location uncertainties <35 km. Eleven more recent major or great earthquakes (magnitude ≥7.7) have been added for completeness through 2004; data from USGS National Earthquake Information Center at http://neic.usgs.gov/ , accessed January 4, 2005. An epicenter is the surface location of the first rupture on an earthquake fault. Symbols shown represent epicenters. For earthquakes larger than about magnitude 7.0, the size of the rupture zone, which can extend hundreds of kilometers from the epicenter, is larger than the symbols used on this map

Depth to earthquake, in km	Magnitude of earthquake			
	5.0–5.9	6.0–6.9	7.0–7.9	≥8.0
<60	•	•	○	○
60–300	•	•	○	○
>300	•	○	○	○
Global average occurrence[1]	1,319/yr	134/yr	17/yr	1/yr

[1]Earthquakes of magnitude <5 (not shown on map) are much more frequent, with ~13,000/yr in the 4.0–4.9 range alone. Data from USGS National Earthquake Information Center.

○ Earthquakes that occurred from 1750 to 1963 within stable plate interiors on continents—Data from A.C. Johnston (Center for Earthquake Research and Information, University of Memphis, written commun., 2002). Even though these epicenters do not meet the precise location criteria of Engdahl and Villaseñor (2002), they are plotted here to remind readers of the potentially hazardous earthquakes that are distant from known plate boundaries. Size of symbol proportional to magnitude of earthquake

○ Notable pre-1900 earthquakes—Nos. 1, 2, 3, 6, and 7 (see table 3, on back)

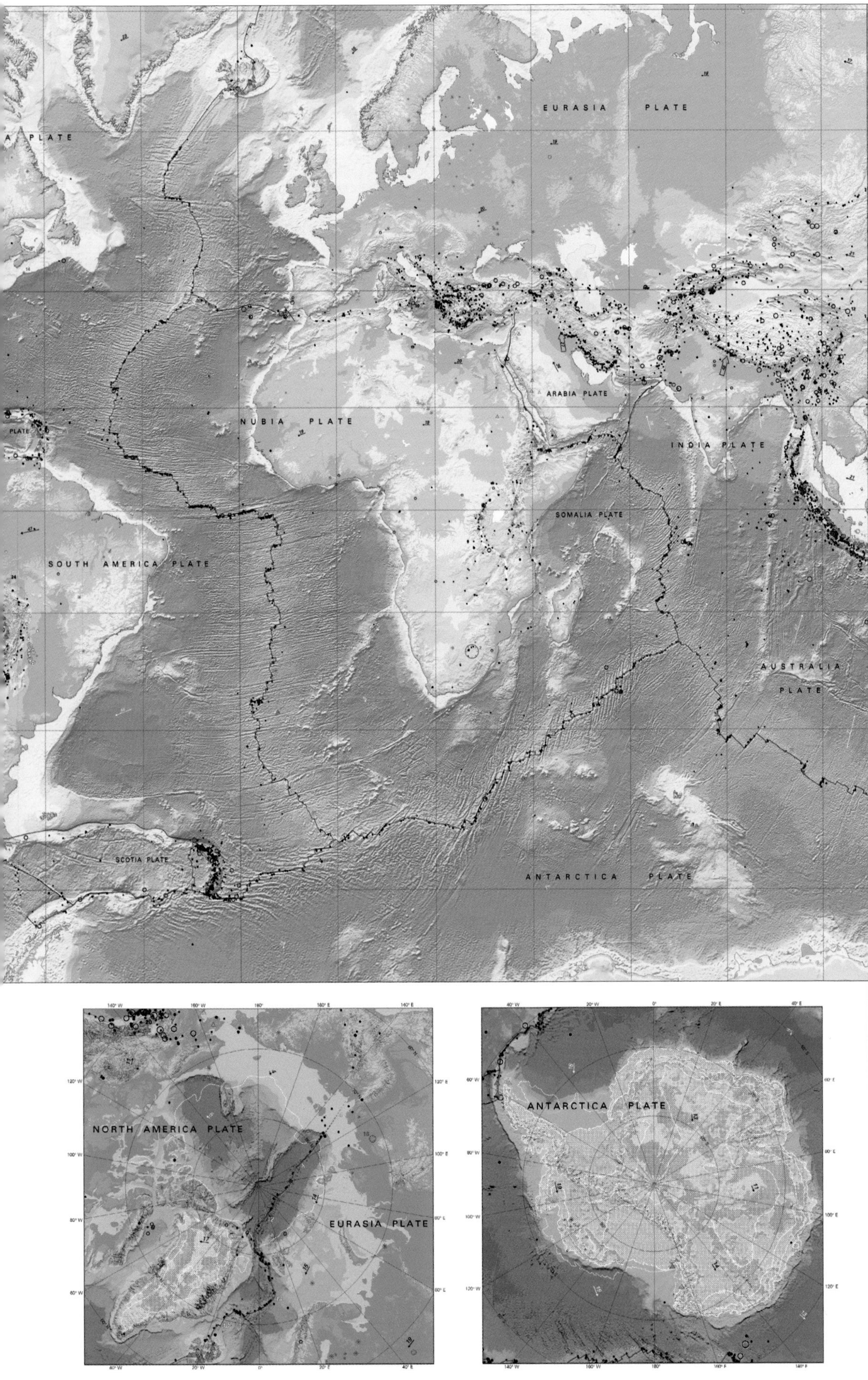
EURASIA PLATE
NUBIA PLATE
ARABIA PLATE
INDIA PLATE
SOMALIA PLATE
SOUTH AMERICA PLATE
AUSTRALIA PLATE
SCOTIA PLATE
ANTARCTICA PLATE
NORTH AMERICA PLATE
EURASIA PLATE
ANTARCTICA PLATE

Creating a 7.5-Minute Topographic Basemap Series

Skagit County GIS
Mount Vernon, Washington, USA
By Joshua D. Greenberg

Contact
Joshua D. Greenberg
joshg@co.skagit.wa.us

Software
ArcGIS Desktop 9.1

Printer
HP Designjet 5000

Data Source(s)
Skagit County, USGS, National Land Cover Database

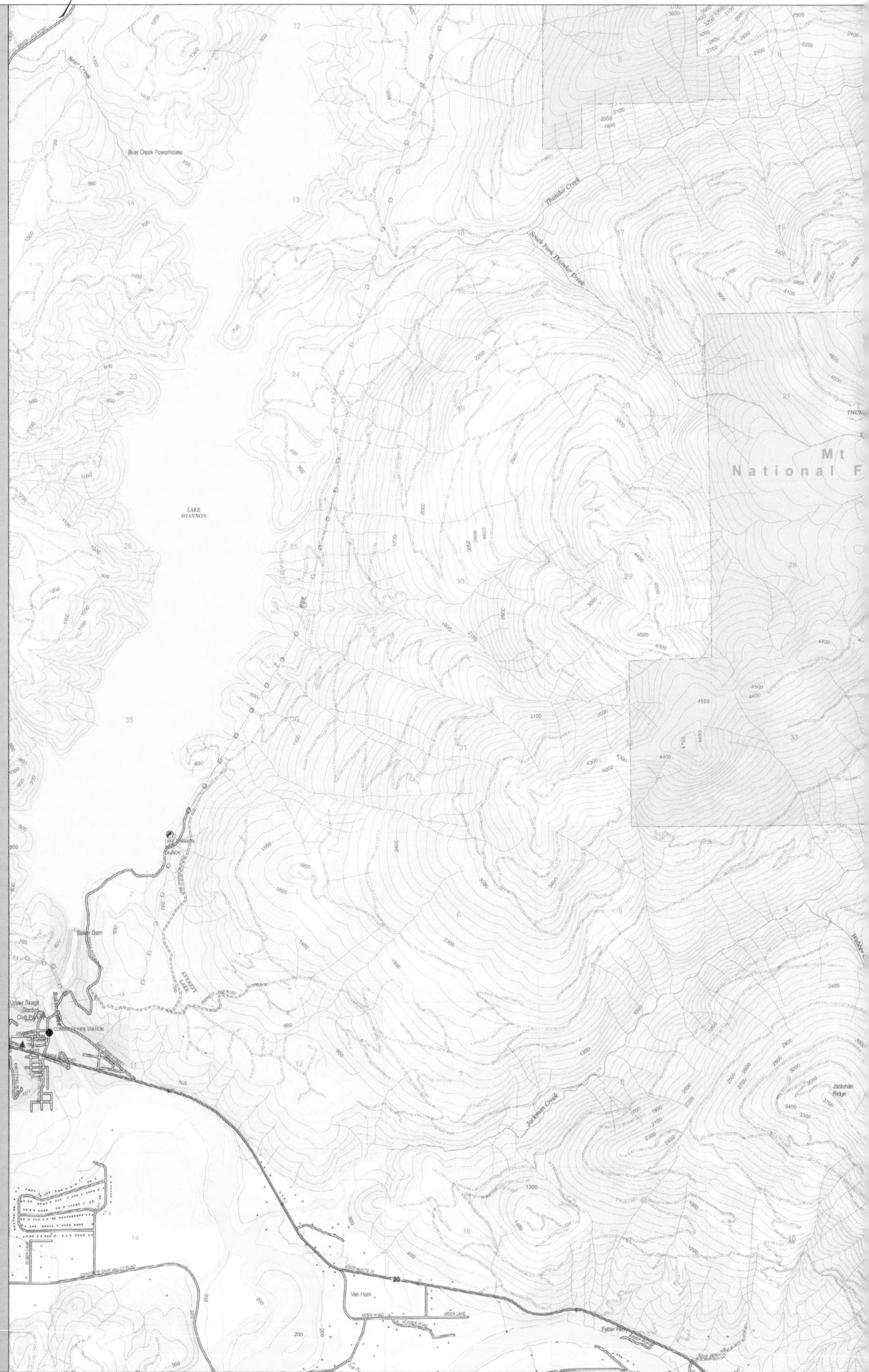

Skagit County departments expressed a desire to have general maps quickly available for search and rescue, emergency response, and basic project planning. The goal for Skagit County GIS was to provide up-to-date maps in a familiar format that had the look and feel of the USGS 7.5-minute quadrangle maps. The maps were created using existing data such as the National Land Cover Database for vegetation, county assessor data for residence locations, and Skagit County's own GIS data for roads and place-names. The surrounding location grid was updated to include the United States National Grid (USNG) coordinates along with latitude and longitude lines. A separate cover sheet contains detailed information such as the legend, explanations of how to use the USNG grid, and disclaimers. There are 40 maps in the series, produced using DS Mapbook and ESRI Maplex extensions for labels. No annotation was used, so the dynamic labeling allows future maps to be created at different scales and extents, and also to be replicated in the county's public ArcIMS site.

Courtesy of Skagit County, Washington.

Joe Leary Slough
Bay View Ridge
Olympia Marsh
Bay View
adilla
Bay
Skagit
Regional
Airport
No Name Slough
Little Indian Slough
TELEGRAPH
SLOUGH
Fredonia
Higgins Slough

Access to Water Trails for the Bay Area

GreenInfo Network
San Francisco, California, USA
By Maegan Leslie Torres

Contact
Larry Orman
larry@greeninfo.org

Software
ArcGIS Desktop 9.1

Printer
HP Designjet 1055cm

Data Source(s)
San Francisco Bay Conservation and Development Commission, GreenInfo Network, ABAG, USGS

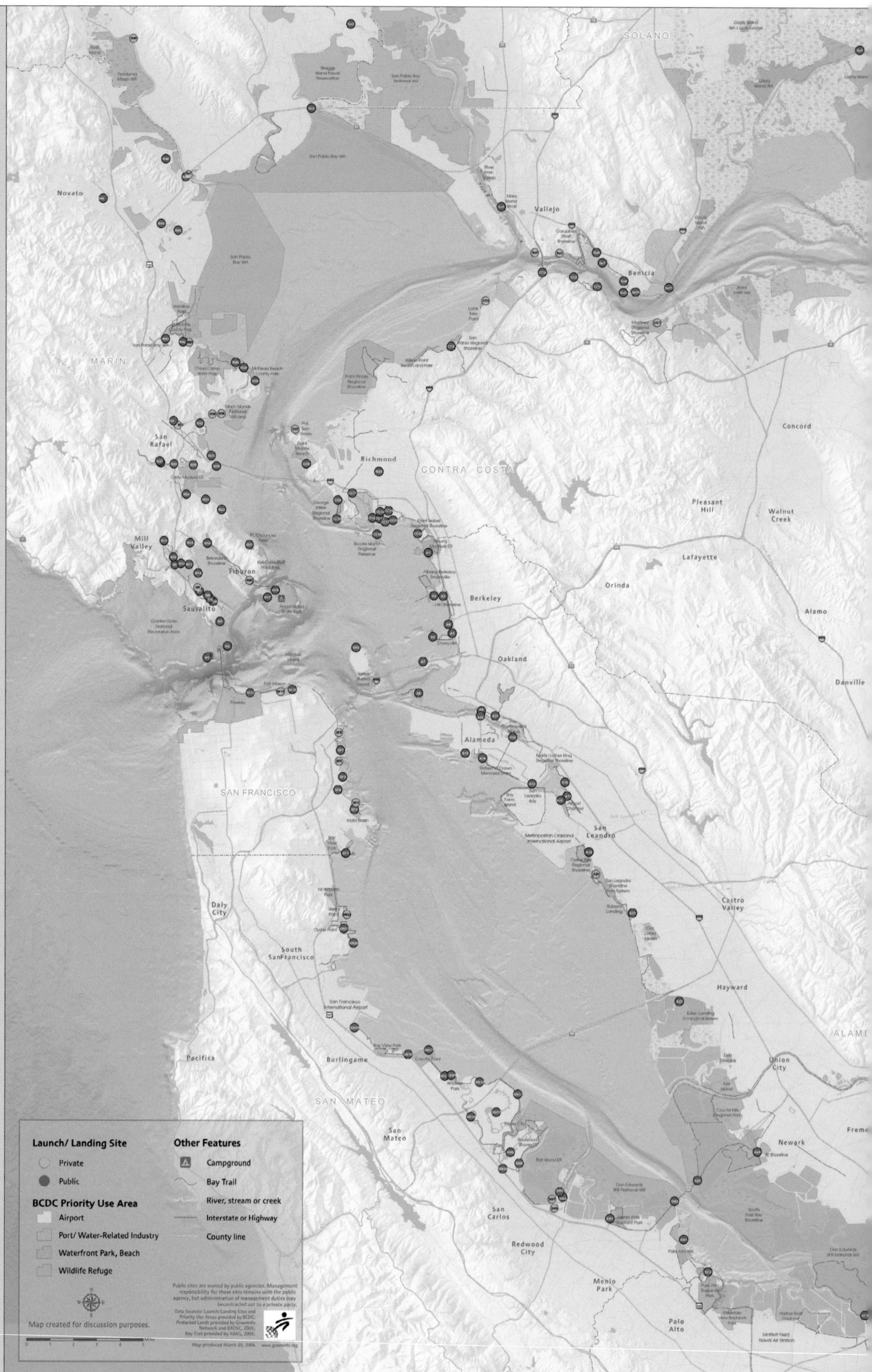

This map was created for the San Francisco Bay Area Water Trail Project, in collaboration with the San Francisco Bay Conservation and Development Commission (BCDC) and the California Coastal Conservancy. It was developed to facilitate planning and discussion at Water Trail Project steering committee meetings. The map focuses on nonmotorized small boat launch and landing sites, BCDC Priority Areas, and other shoreline recreational uses. The bathymetry of the bay was used to draw the viewer away from the surrounding land and to focus on the water and shoreline.

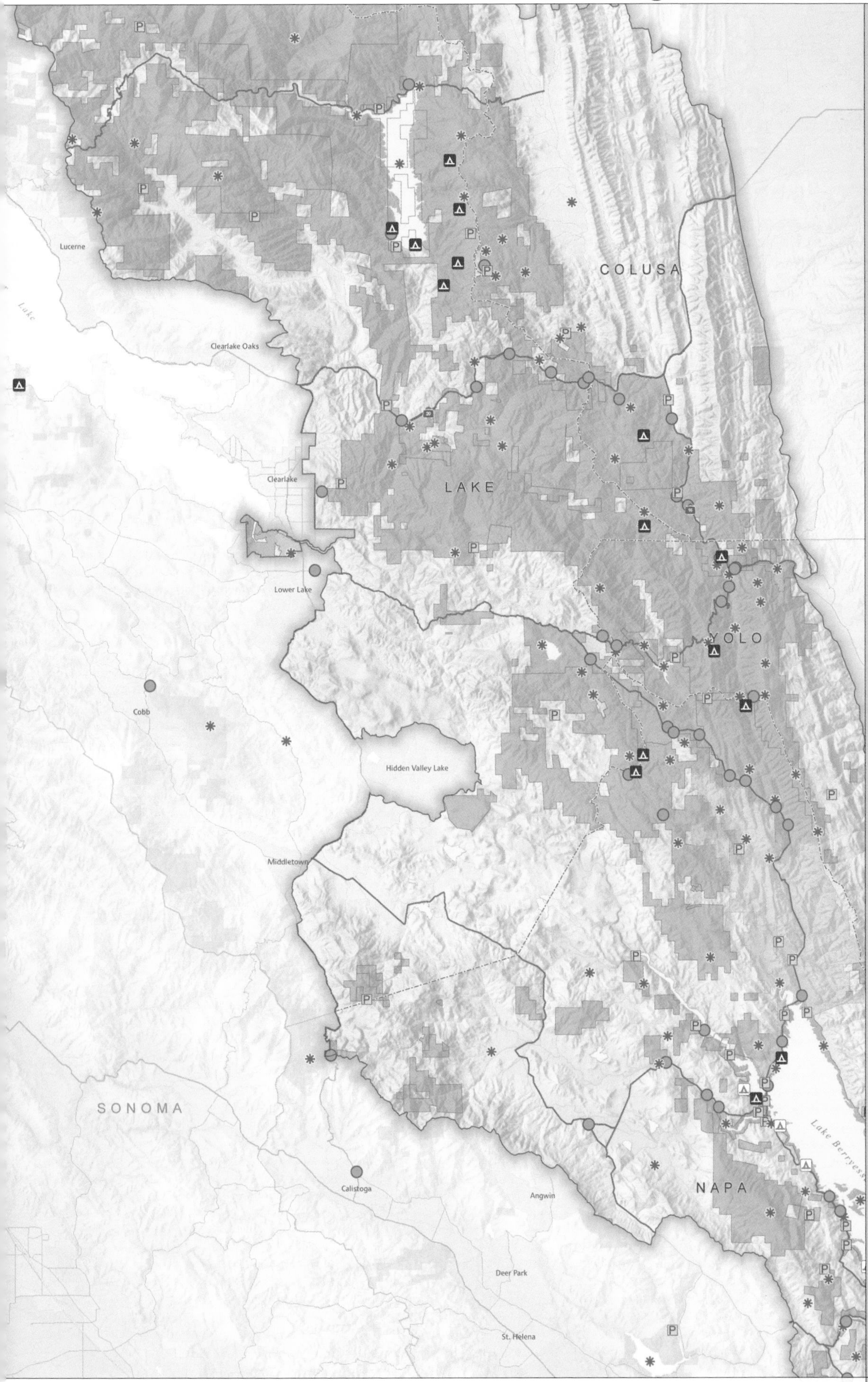

GreenInfo Network
San Francisco, California, USA
By Maegan Leslie Torres

Contact
Larry Orman
larry@greeninfo.org

Software
ArcGIS Desktop 9.1

Printer
HP Designjet 1055cm

Data Source(s)
GreenInfo Network, BRNBA Conservation Partnership, Jake Mann masters thesis, USGS

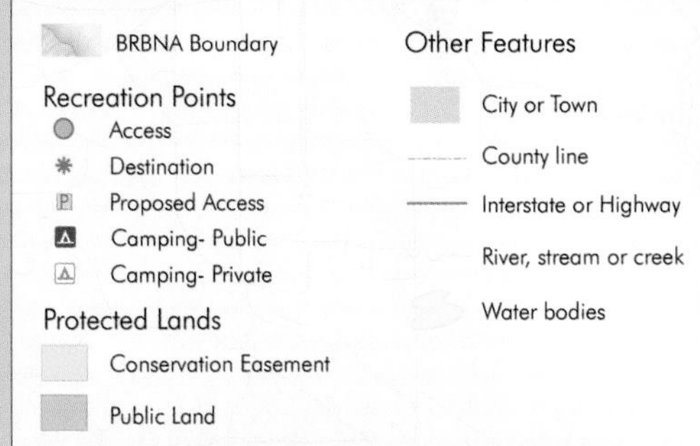

Created for the Blue Ridge Berryessa Natural Area (BRBNA) Conservation Partnership, this map is intended to showcase the recreational priorities in the BRBNA, and to serve as a planning tool for the partnership. The BRBNA is located less than two hours from California's Sacramento and San Francisco Bay areas and includes the rugged, natural landscapes and rangelands of the Putah and Cache creeks watersheds. The area supports a unique assemblage of ecological communities, abundant and diverse wildlife, and indigenous plants. The biodiversity and scenic beauty of the area creates unique recreational opportunities, highlighted in the map. In order to bring out the BRBNA, a transparent mask was used on the surrounding areas and a transparent feathered buffer was added to highlight the region.

Applied Geographics, Inc.

Boston, Massachusetts, USA
By Applied Geographics, Inc.

Contact

David Weaver
weaver@appgeo.com

Software

ArcGIS Desktop

Data Source(s)

Interviews and field source data

Acquired Data Sets

- CRUISE_SHIPPING
- DREDGE_AREA_PTS
- GAMBLING_BOATS
- KAYAK_LAUNCH_SITES
- LARGEMOTOR_REGISTERED
- MARINAS
- RECREATIONAL_DIVE_SITES
- SAILBOATS_REGISTERED
- SEAPORTS
- WATER_TAXI_STOPS
- WHALEWATCH_BERTHS

Whale Sitings, 1962-2006

- Beaked North Sea Whale
- Beluga - White Whale
- Blainville's Beaked Whale
- Blue Whale
- Dwarf Sperm Whale
- False Killer Whale
- Fin Whale
- Goose-beaked Whale
- Harbor Porpoise
- Humpback Whale
- Killer Whale
- Minke Whale
- Northern Bottlenose Whale
- Northern Right Whale
- Pygmy Killer Whale
- Pygmy Sperm Whale
- Sei Whale
- Sperm Whale
- True's Beaked Whale

Dredging Spoils Sites

- Active
- Historical
- Under review

- FERRY_ROUTES
- DREDGE_AREAS
- BEACH_NOURISHMENT_AREAS
- MOORING_FIELDS
- KAYAK_RECREATION_AREAS
- DESALINATION_FACILITIES
- ENERGY_FACILITIES

Large Sailboat Storage by Town (count)

- 0 - 21
- 22 - 66
- 67 - 141
- 142 - 220
- 221 - 363

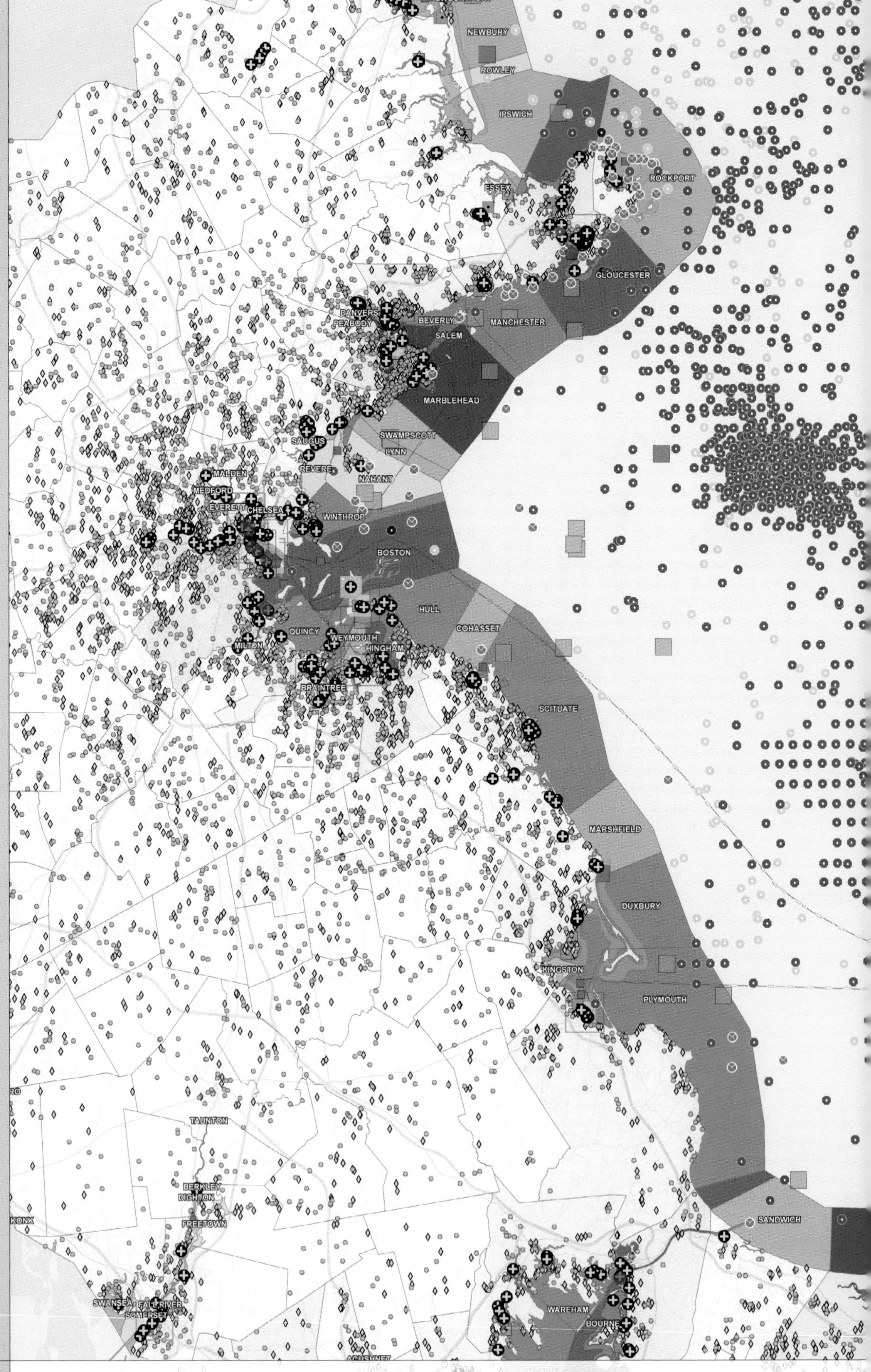

Applied Geographics, Inc., was contracted to provide the Massachusetts Office of Coastal Zone Management (MCZM) with GIS data research, collection, automation, standardization, and metadata development services. The datasets all relate to human uses of the coastal zone. About 50 types of data were researched and documented, and about 20 were prioritized for collection and conversion to MCZM standards for delivery. The human-use GIS data included dredging and dredge-spoil areas, cruise-ship berthing, motor and sailboat registrations, marinas, yacht clubs, mooring areas, dredge-spoil sites, water-taxi stops, and recreational dive sites.

Courtesy of Applied Geographics, Inc.

Ecological Characterization of the Stellwagen Bank National Marine Sanctuary Region

National Oceanic and Atmospheric Administration
Silver Spring, Maryland, USA
By B. Costa, F. Huettmann, S. Pittman, T. Battista, and K. Eschelbach

Contact
Bryan Costa
bryan.costa@noaa.gov

Software
ArcGIS Desktop 9.1

Printer
HP Designjet 5500

Data Source(s)
Manomet Bird Observatory, SeaWiFS and MODIS remotely sensed imagery, USGS bathymetry

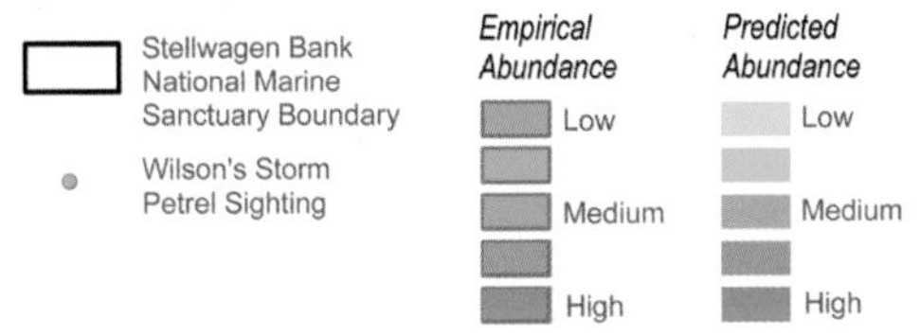

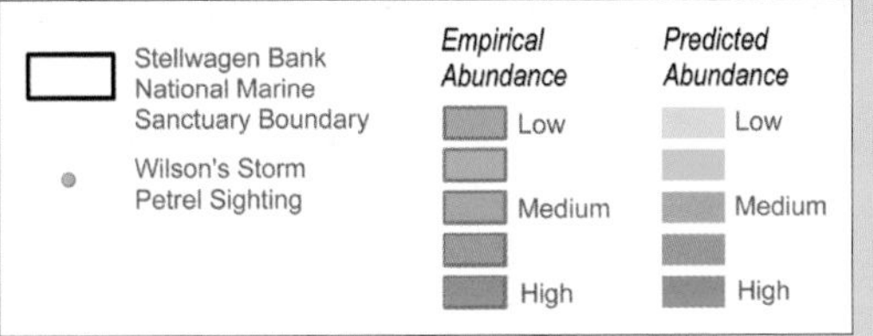

Biogeographic assessments enable scientists and resource managers to evaluate biodiversity across multiple spatial and temporal scales. This scalar understanding empowers administrators to make informed decisions in support of science-based ecosystem management. The Center for Coastal Monitoring and Assessment's Biogeography Team, in collaboration with the National Marine Sanctuary Program, conducted an ecological characterization of Stellwagen Bank National Marine Sanctuary (SBNMS) and the surrounding marine region. The purpose of this project was to enhance the understanding of key ecological patterns and processes in the Gulf of Maine, in order to assist the SBNMS management plan review process.

To identify these patterns and processes, an interdisciplinary team of scientists collected, integrated, and analyzed a wide range of *in situ* and remotely sensed physical, chemical, and biological oceanography data in ArcGIS. Statistical and spatial analytical techniques were used to describe spatiotemporal patterns in the distribution of contaminants, pelagic invertebrates, fish, seabirds, and marine mammals. Predictive models offered promising opportunities to extrapolate discrete regional information to broad biogeographic scales, where spatial gaps in the data exist.

Courtesy of National Oceanic and Atmospheric Administration.

CommEn Space
Seattle, Washington, USA
By Matthew Robert Stevenson and CommEn Space

Contact
Matthew Robert Stevenson
matt@coregis.net

Software
ArcGIS Desktop 9.1, Adobe Photoshop

Printer
HP Designjet 500

Data Source(s)
USGS, Washington Natural Heritage Program, Washington Department of Natural Resources; Washington Department of Fish and Wildlife; Oregon Heritage Information Center

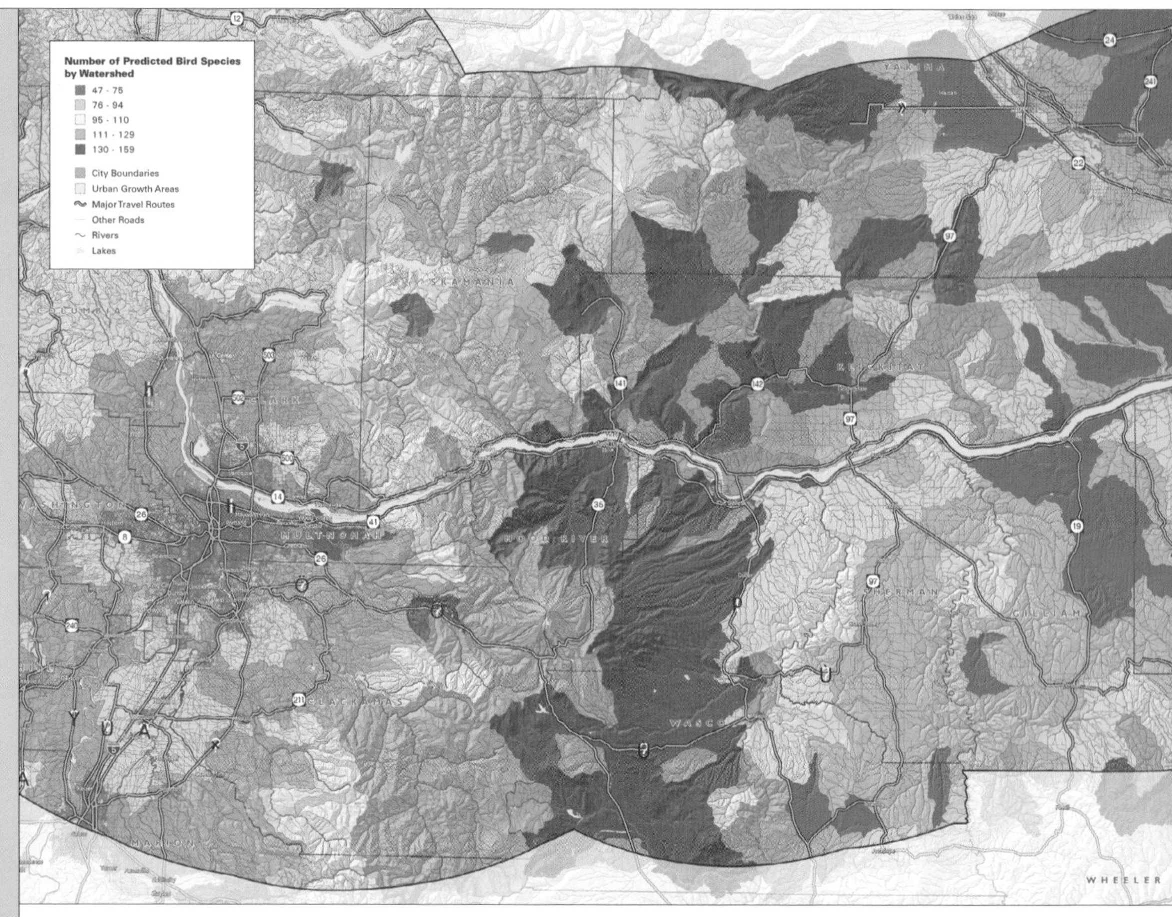

This Service Area-Wide Comparison map, derived from Washington and Oregon Gap Analysis data, shows areas of high bird-species richness. The predicted distributions for all birds in both states were summarized within each watershed falling within the Columbia Land Trust study area. The results are displayed using a natural breaks algorithm to show naturally occurring patterns in the data. Watersheds colored dark greenish-blue have the highest number of predicted species, and watersheds colored brown have the lowest.

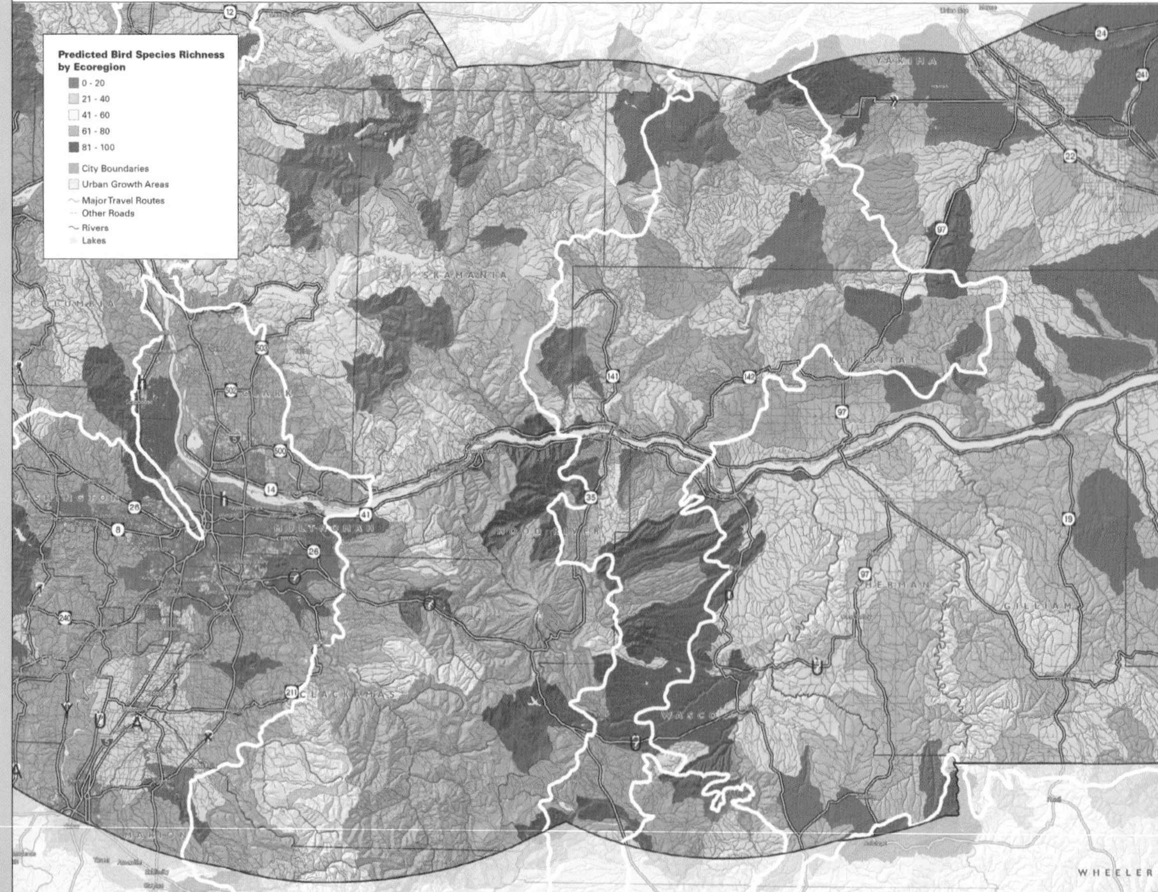

In the Ecoregional Comparison of Bird Species Richness map, the results for each watershed were scaled within each of the Columbia Land Trust service area's seven ecoregions. For each ecoregion, the watershed containing the highest number of predicted bird species was set to "100" and the watershed containing the lowest number was set to "0." All other values in between were scaled between 0 and 100 to yield a score. The end result is that watersheds in the Coast Range ecoregion, for example, are not compared with watersheds in the Columbia Plateau ecoregion.

Courtesy of CommEn Space.

CommEn Space

Seattle, Washington, USA
By Matthew Robert Stevenson and CommEn Space

Contact

Matthew Robert Stevenson
matt@coregis.net

Software

ArcGIS Desktop 9.1, Adobe Photoshop

Printer

HP Designjet 500

Data Source(s)

USGS, Washington Natural Heritage Program, Washington Department of Natural Resources, Oregon Heritage Information Center

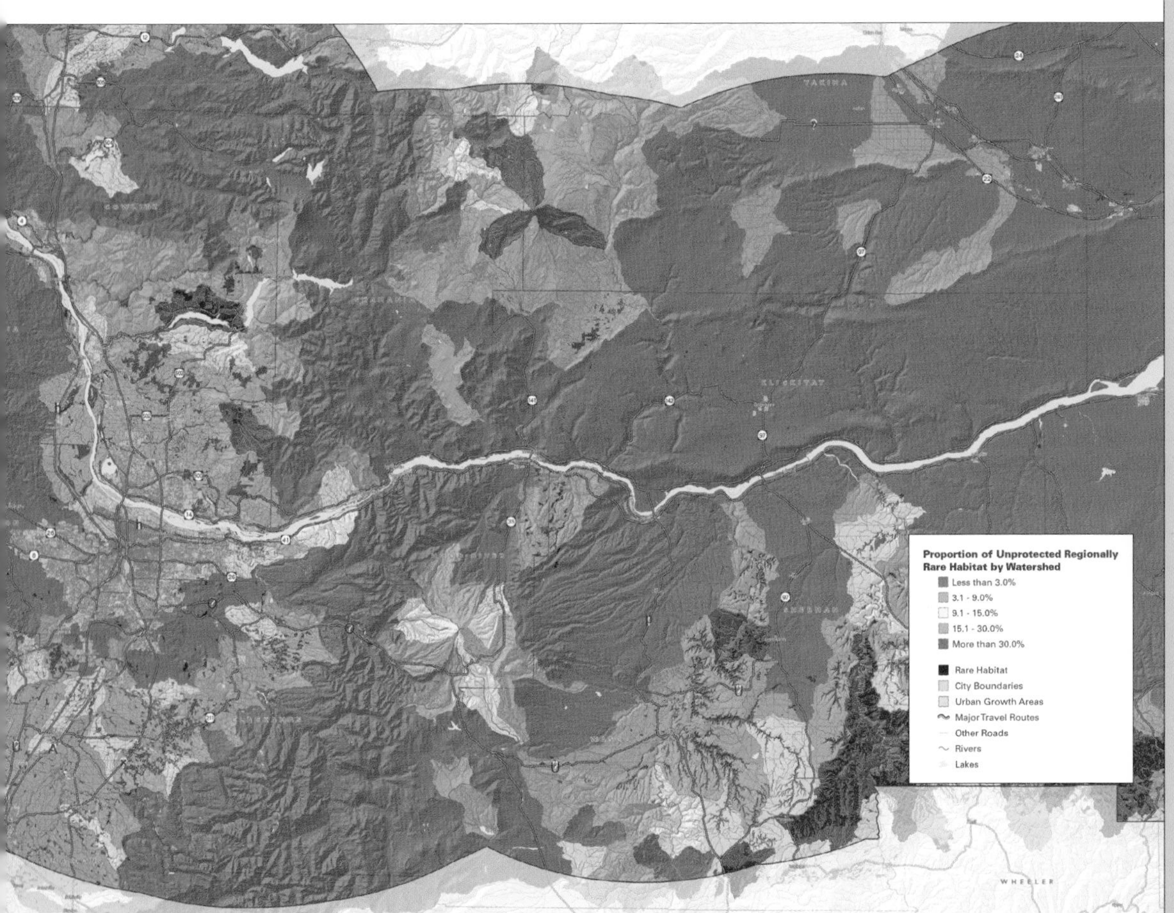

The Columbia Land Trust is currently updating its five-year conservation plan with the assistance of CommEn Space. The plan incorporates data on species distribution, habitat types, protected lands, and development threat, and will be used by the land trust to prioritize its conservation activities. Using data produced by the Northwest Habitat Institute, CommEn Space analyzed the extent and conservation status of the 24 habitat types located within the Columbia Land Trust study area. Eleven habitat types cover less than 0.5 percent of the study area, and an additional five cover between 0.5 and 1.0 percent. This map shows which watersheds contain the most unprotected rare habitat based on the sum of the proportion of unprotected habitat across all rare habitat types. This map does not show the variety of rare habitat types within each watershed.

Courtesy of CommEn Space.

Cruise Ship Overboard Discharge Project

Conservation International
Washington, D.C., USA
By Erica Ashkenazi

Contact

Mark Denil
m.denil@conservation.org

Software
ArcGIS Desktop 9.1

Printer
HP Designjet 5500ps

Data Source(s)
Conservation International, National Imagery and Mapping Agency (NIMA)

- Cities
- Discharge Start Points with Vector
- Land
- Marine Protected Area Polygons
- Marine Protected Area Points
- 12 Nautical Miles from Coast
- Coral Reefs
- Bathymetric Contours (m)
  - 60
  - 100

In order to estimate general patterns of cruise-line waste discharge in time and space, the Ocean Conservation and Tourism Alliance (OCTA) Science Panel requested data on discharges directly from the International Council of Cruise Lines (ICCL). The start and stop point for each discharge event was recorded in the database. It was assumed that each ship traveled in a straight line following the shortest-path distance on the globe during its time of discharge. For each pair of recorded start and stop points, a line feature was created in a gnomonic projection to produce an arc along the corresponding great circle for each point pair. These lines were then unprojected into decimal degrees for subsequent analysis.

To help eliminate data entry errors, modeled line segments that intersected any part of a land feature were eliminated from subsequent analyses. The coastline used for determining land features was NIMA's VMap Level 0, a GIS dataset of base-data features at 1:1,000,000 scale. To analyze geographic patterns in the distribution of waste, summary statistics were calculated using a 0.5-degree grid cell system. This returned the total count of lines per 0.5 x 0.5-degree cell for various subsets of the data.

Courtesy of Conservation International.

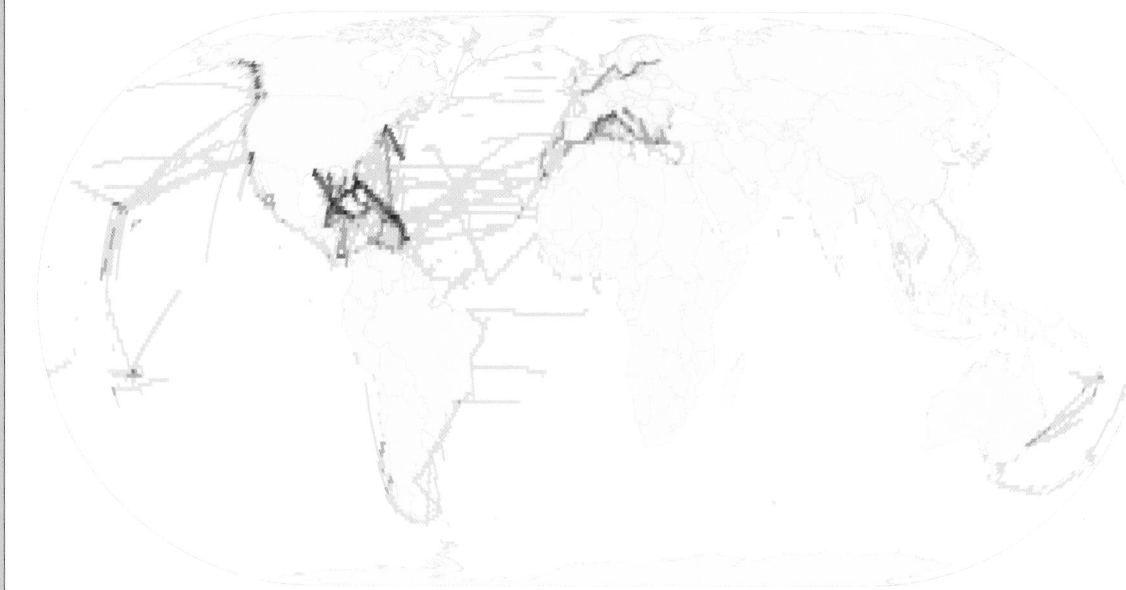

Density of Discharge Tracks

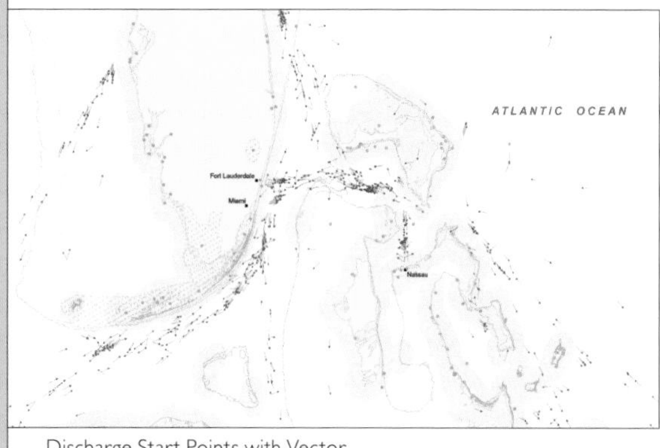

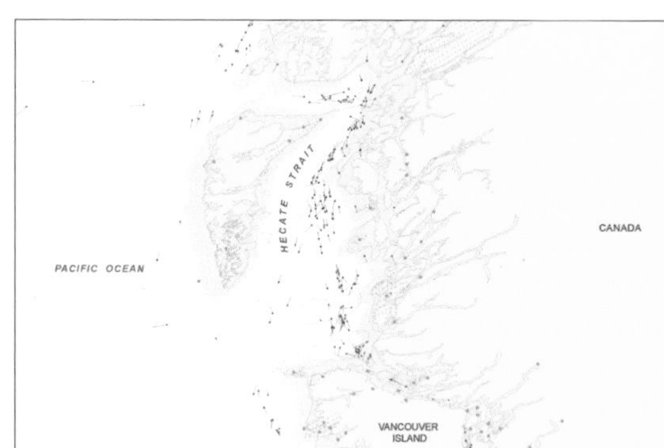

Discharge Start Points with Vector

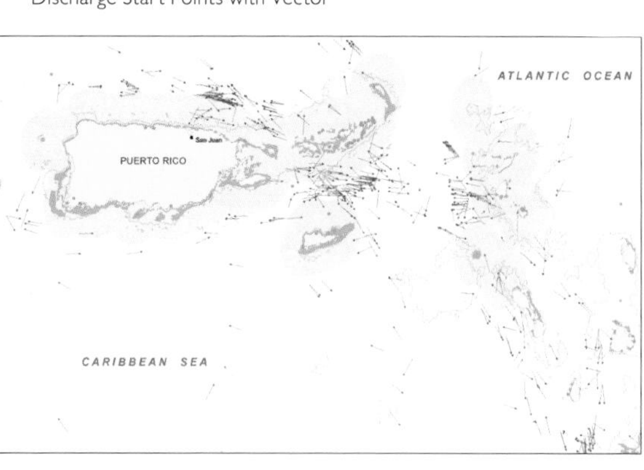

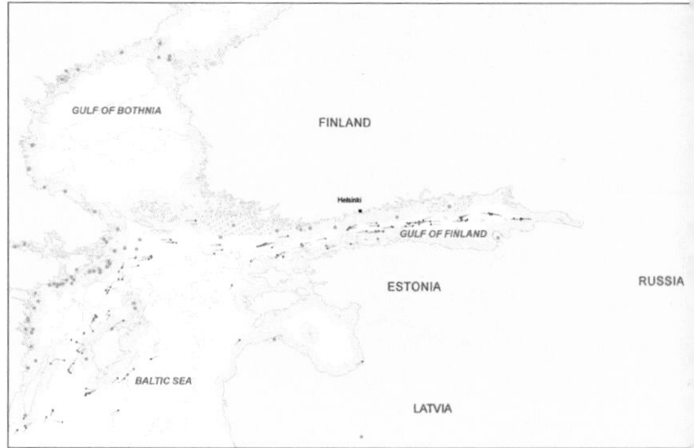

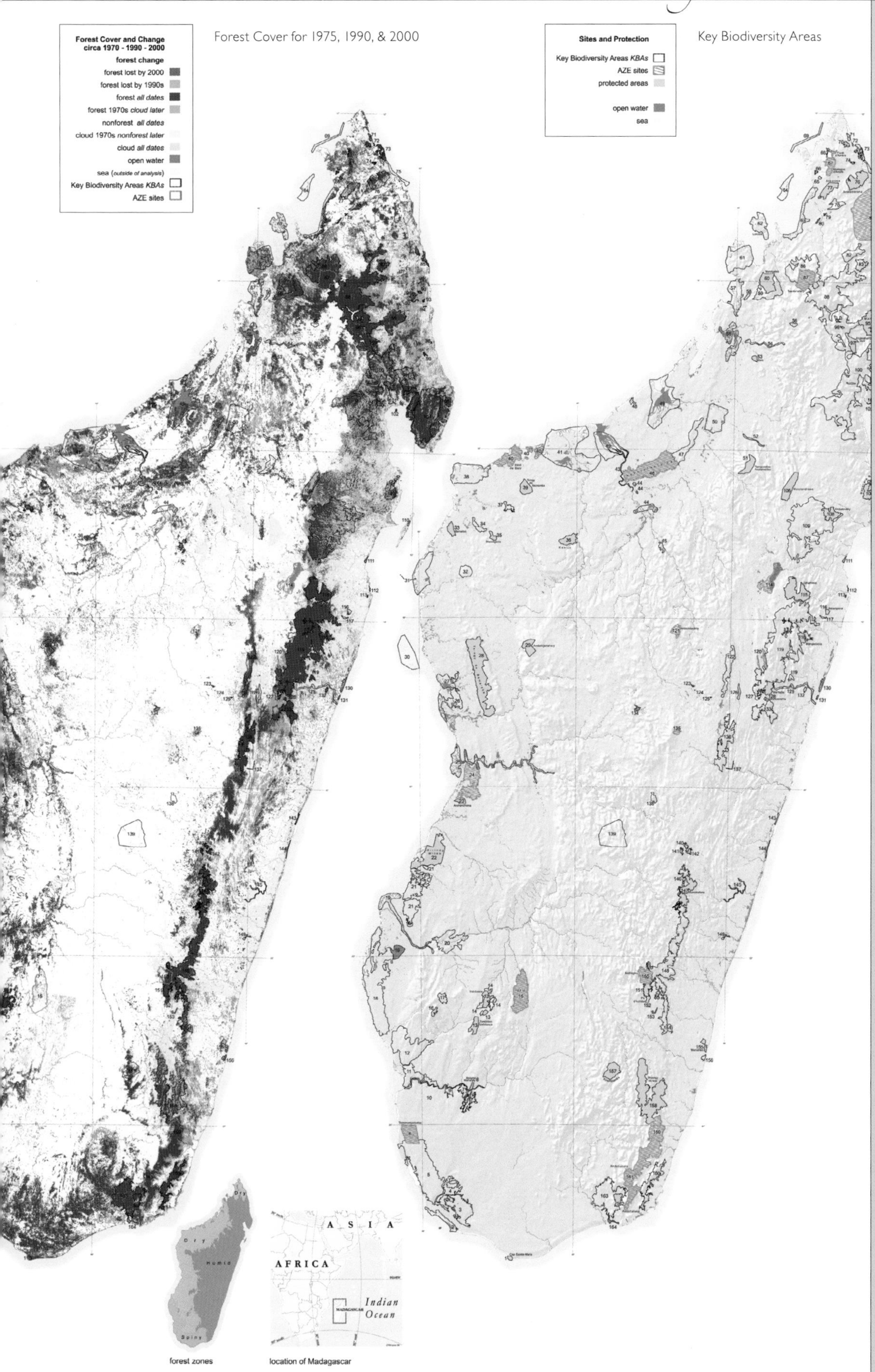

Conservation International
Washington, D.C., USA
By Mark Denil

Contact
Mark Denil
m.denil@conservation.org

Software
ArcGIS Desktop 9.1

Printer
HP Designjet 5500ps

Data Source(s)
Conservation International

One of the major knowledge gaps facing conservationists and policy makers has been the lack of reliable information about ecosystem health and rates of landscape change in both terrestrial and marine environments. The mission of The Center for Applied Biodiversity Science's Regional Analysis Program at Conservation International is to use satellite, aerial, and field observations to characterize and monitor the impacts of human activities on biodiversity in the hot spots. Integrating this new generation of space and airborne remote sensing instrumentation with comprehensive databases on social, economic, political, and legal factors enables better understanding of the relationships between the biophysical environment and patterns of human use.

Key biodiversity areas are places of international importance for conservation via protected areas and other governance mechanisms. They are identified nationally using simple standard criteria, based on their importance in maintaining populations of species. Key biodiversity areas are the building blocks for designing the ecosystem approach and maintaining effective ecological networks—the starting point for landscape-level conservation planning.

Courtesy of Conservation International.

Arctic Conservation Area Topographic Map

UNEP/GRID-Arendal
Stockholm, Sweden
By Hugo Ahlenius

Contact
Hugo Ahlenius
hugo@grida.no

Software
ArcGIS Desktop 9.1, Adobe Illustrator CS2, Adobe Photoshop CS2

Data Source(s)
Conservation of Arctic Flora and Fauna Working Group of the Arctic Council

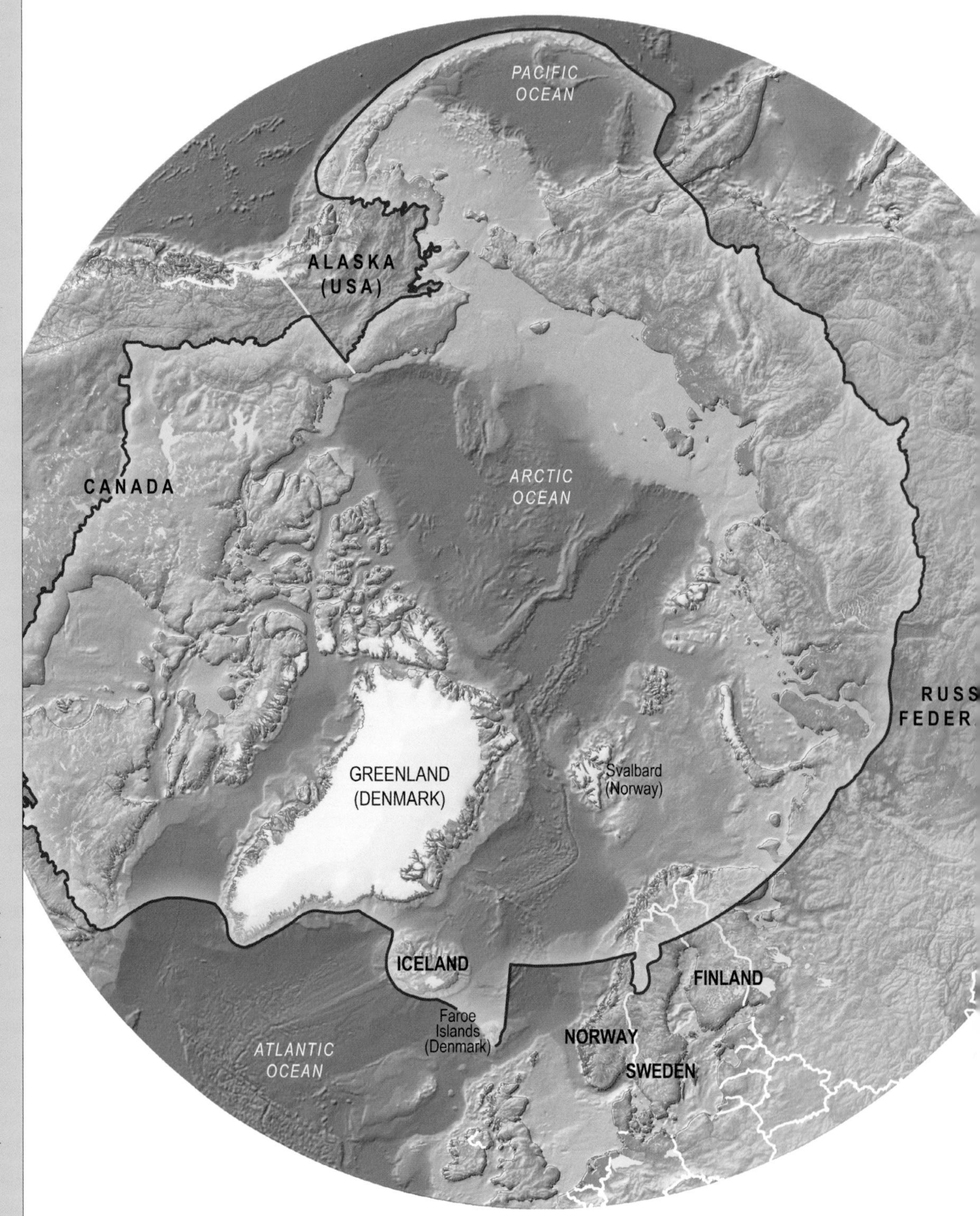

The Conservation of Arctic Flora and Fauna (CAFF) is a working group under the Arctic Council for the countries of the Arctic, indigenous peoples, and nongovernmental organizations. Monitoring, assessment, protection of areas, and conservation strategies are all tasks under this working group. The area that this group addresses is represented in this map.

The vast Arctic area represents one of the last wilderness regions in the world, with very little human activity. The boreal forest, tundra, and glaciated areas of the Arctic are home to polar bears, caribou, and many other similar species. For many migrating birds, this is the destination for their summer breeding.

The Arctic faces considerable challenges in the future, in the continued development of the extraction of natural resources such as oil and gas, climate change, and preservation of the traditional lives of the indigenous peoples. The changes in climate that can already be seen are projected to have the most significant impact in this part of the world, with less sea ice and the melting of permafrost in the tundra.

Courtesy of Hugo Ahlenius, UNEP/GRID-Arendal.

University of Redlands
Redlands, California, USA
By Stephen Norris

Contact
Stephen Norris
stevesan74@gmail.com
msgis@redlands.edu

Software
ArcGIS Desktop, Adobe Illustrator, SketchUp

Printer
HP Designjet 1055cm

Data Source(s)
ESRI, USGS, Azsite, Professor Wesley Bernardini

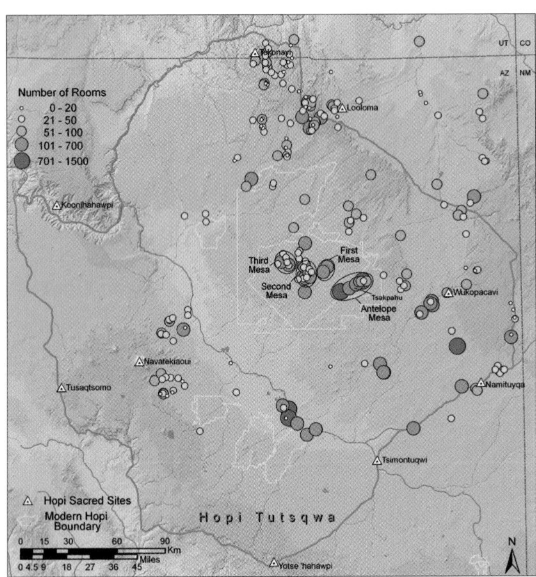

Archaeological Hopi sites

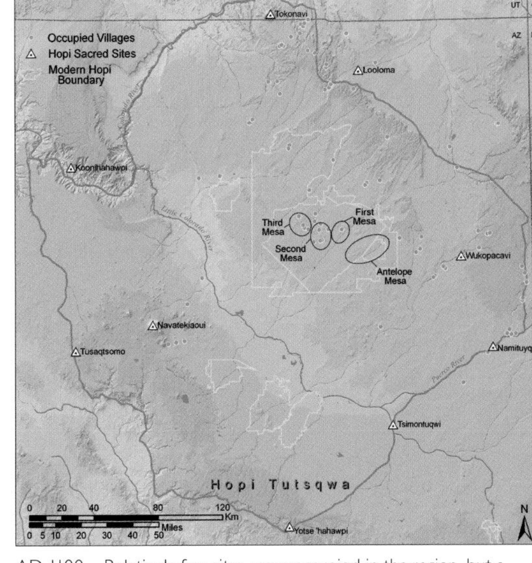

AD 1100—Relatively few sites were occupied in the region, but a large number of sites ring its periphery.

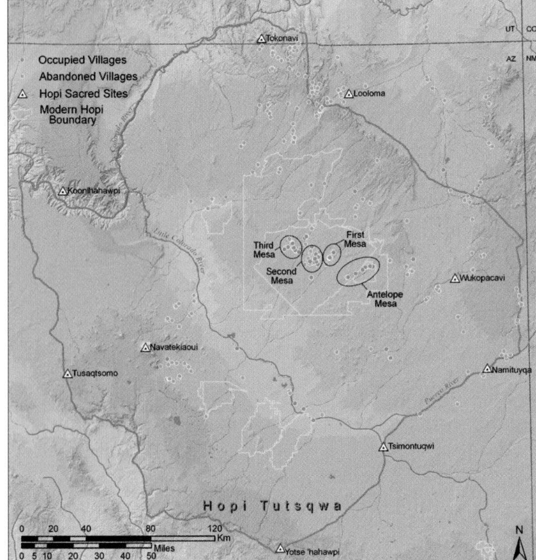

AD 1300—Many of the peripheral areas had been abandoned and many of their occupants had moved into the Hopi region to found new villages.

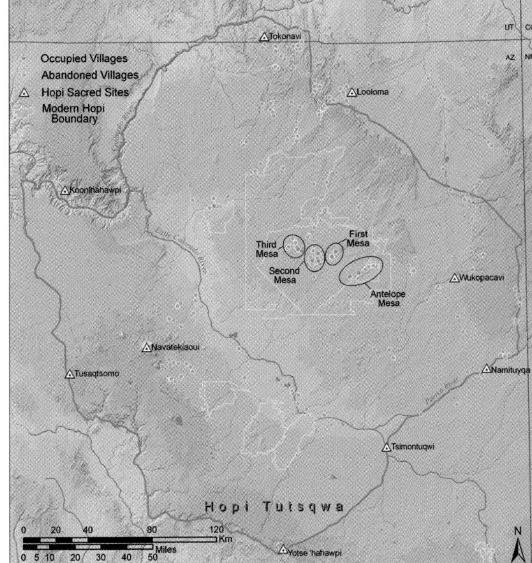

AD 1500—The population in the Hopi region had consolidated into fewer, larger villages, leaving a core settled area surrounded by a large tract of unoccupied land.

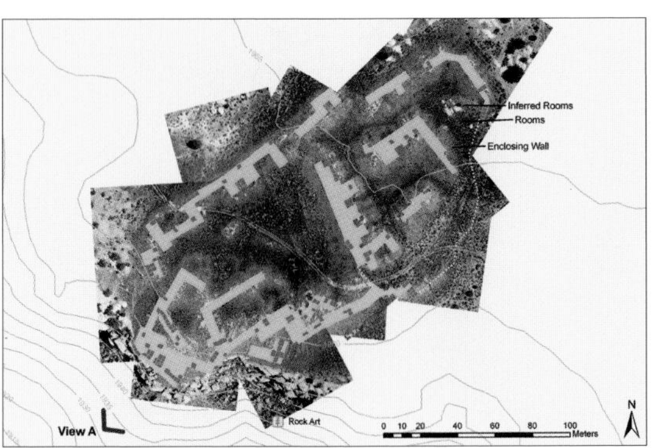

Tsakpahu village

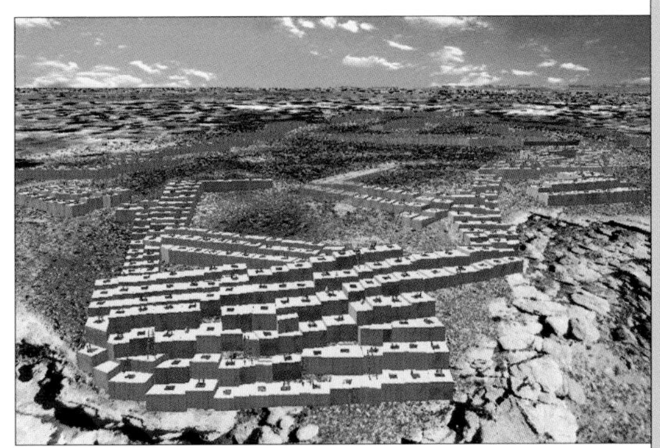

3D model of Tsakpahu village

The University of Redlands and the Hopi Tribe have collaborated to identify ancestral sites and to reconstruct clan migration patterns. Professor Wesley Bernardini, PhD, defined the primary goals for the project: to develop a GIS database for storage and analysis of spatial data on ancestral Hopi villages, to use travel times to evaluate relationships among villages, and to build 3D reconstructions of individual villages.

Cost Path Analysis

A Cost Path Analysis calculates travel time that accounts for the difficulty of crossing different kinds of terrain. The Cost Path displays a line of travel that accounts for the least cost based on a Cost Surface input. This Cost Surface is a reclassification of percentage of slope into time, in the seconds it takes to cross a given cell (30 meters in this case). The Cost Distance Surface is the accumulation of time radiating out from a specified location.

3D Visualization

Georeferenced room and building footprints were imported into ArcScene software along with USGS elevation data and Terraserver satellite imagery (view A) to create a 3D visualization of a Hopi village. Balloon photos were draped over the elevation (view A, bottom left) to provide more realistic details. The polygons that make up the rooms were extruded in the layer properties and enhanced in SketchUp, giving it a three-dimensional representation.

Time

Plotting the sequence of ancestral Hopi villages as they were founded and abandoned was the first step in demographic reconstruction toward understanding how populations came together to form the Hopi. A village database was used to map the locations of villages over time. Part of the archaeological information consisted of periods of human occupation for each village: AD 1100, AD 1300, and AD 1500. These results show the growth and movement of the Hopi population using three sequential time snapshots.

Courtesy of Professor Wesley Bernardini, PhD.

Missing Infrastructure by Elementary School Area

City of Chula Vista

Chula Vista, California, USA

By Jack Hurlbut

Contact

Jack Hurlbut

jhurlbut@ci.chula-vista.ca.us

Software

ArcGIS Desktop, @City iStreetView

Printer

HP Designjet 4000

Data Source(s)

City datasets (ArcSDE, orthophotos), street-level photos by @City Inc.

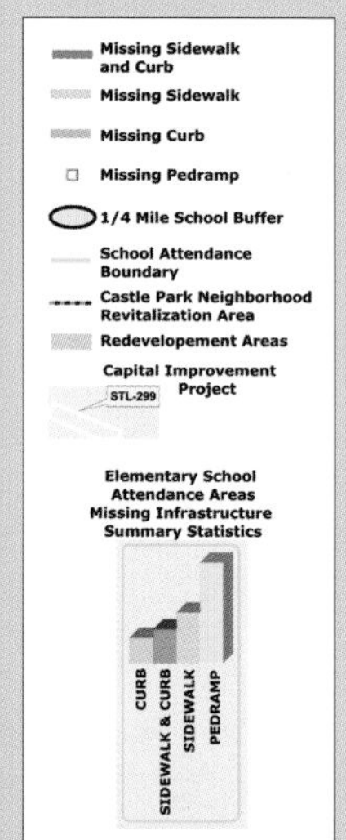

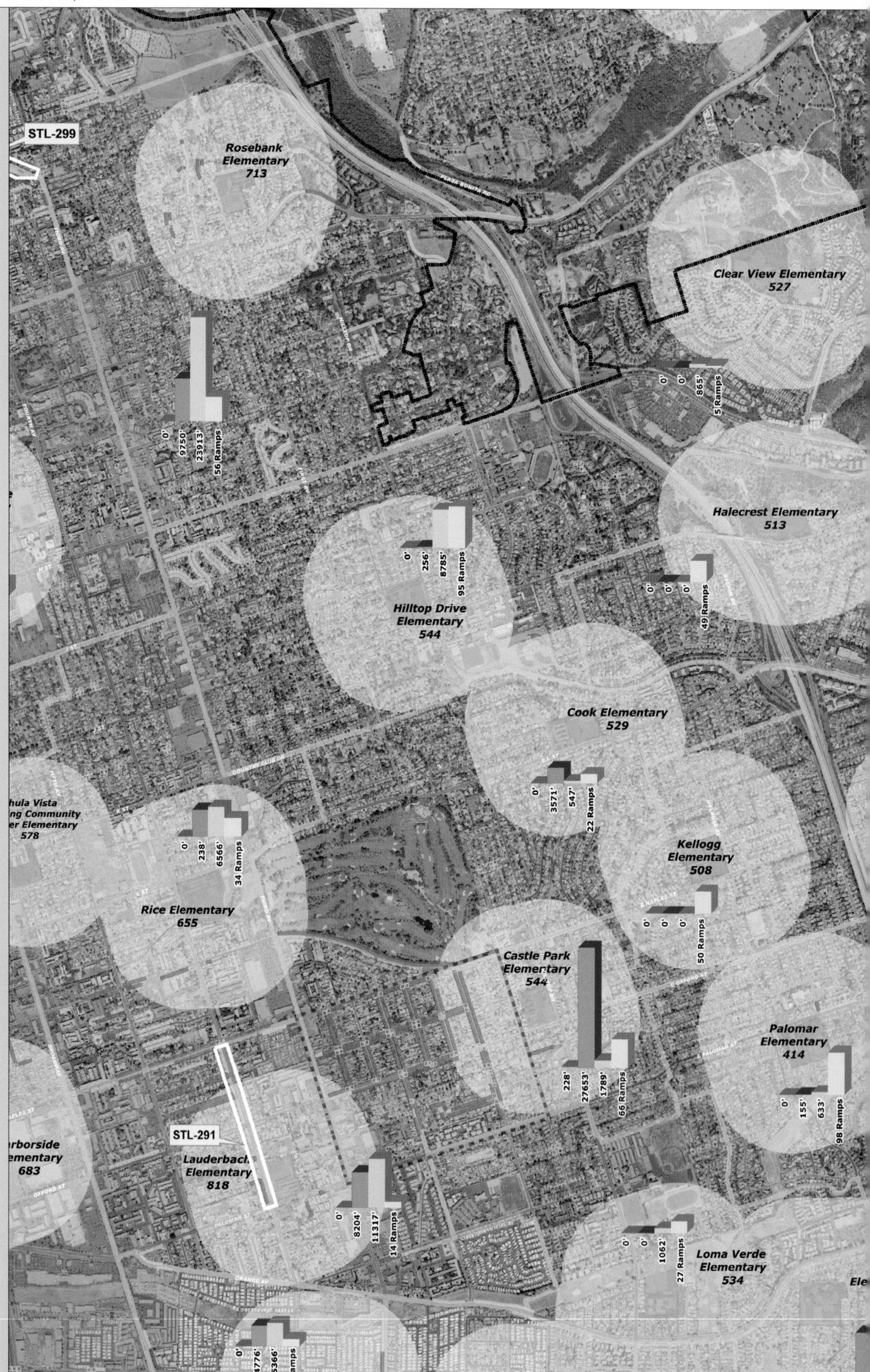

Chula Vista is located in southern San Diego County and encompasses approximately 52 square miles. The city has the eighth-fastest percentage growth rate nationally since the 2000 Census. The Chula Vista Elementary School District is the largest K–6 district in the state of California. The map shown here was produced for the Engineering Department's Infrastructure Services Division. It is responsible for identifying current and future deficiencies in the city's pavement, storm-drain, wastewater-disposal, and street systems.

To aid in its undertaking, the Engineering Department relies upon the city's GIS Division of the Information and Technology Services Department. There are more than 100 layers in the GIS, encompassing the entire spectrum of municipal needs. Of special importance to the Missing Infrastructure project were the street-level geocoded digital images of all city streets that Chula Vista had @City Inc. take. Use of @City Inc.'s iStreetView software to examine these images while using ArcMap 9.1 software to view orthophotos and GIS layers such as sidewalk, pedramp, and street, sped up the detection of missing infrastructure by virtually eliminating field inspections. It was felt that a community-focused division based on elementary school attendance areas would provide a convenient size allowing all interested parties to discuss priorities. Using symbology and chart tools in ArcMap software, point and linear information was presented as bar charts for each school's attendance area. This graphic enhancement provided a clear indication of where resources needed to be allocated.

Courtesy of the City of Chula Vista, California.

NASA Langley Research Center, GIS Team
Langley, Virginia, USA
By Sarah K. Smith and Dana Torres

Contact
Sarah K. Smith, Dana Torres
s.k.smith@larc.nasa.gov, d.c.torres@larc.nasa.gov

Software
ArcGIS Desktop 9.1

Data source(s)
NASA LaRC, USDA

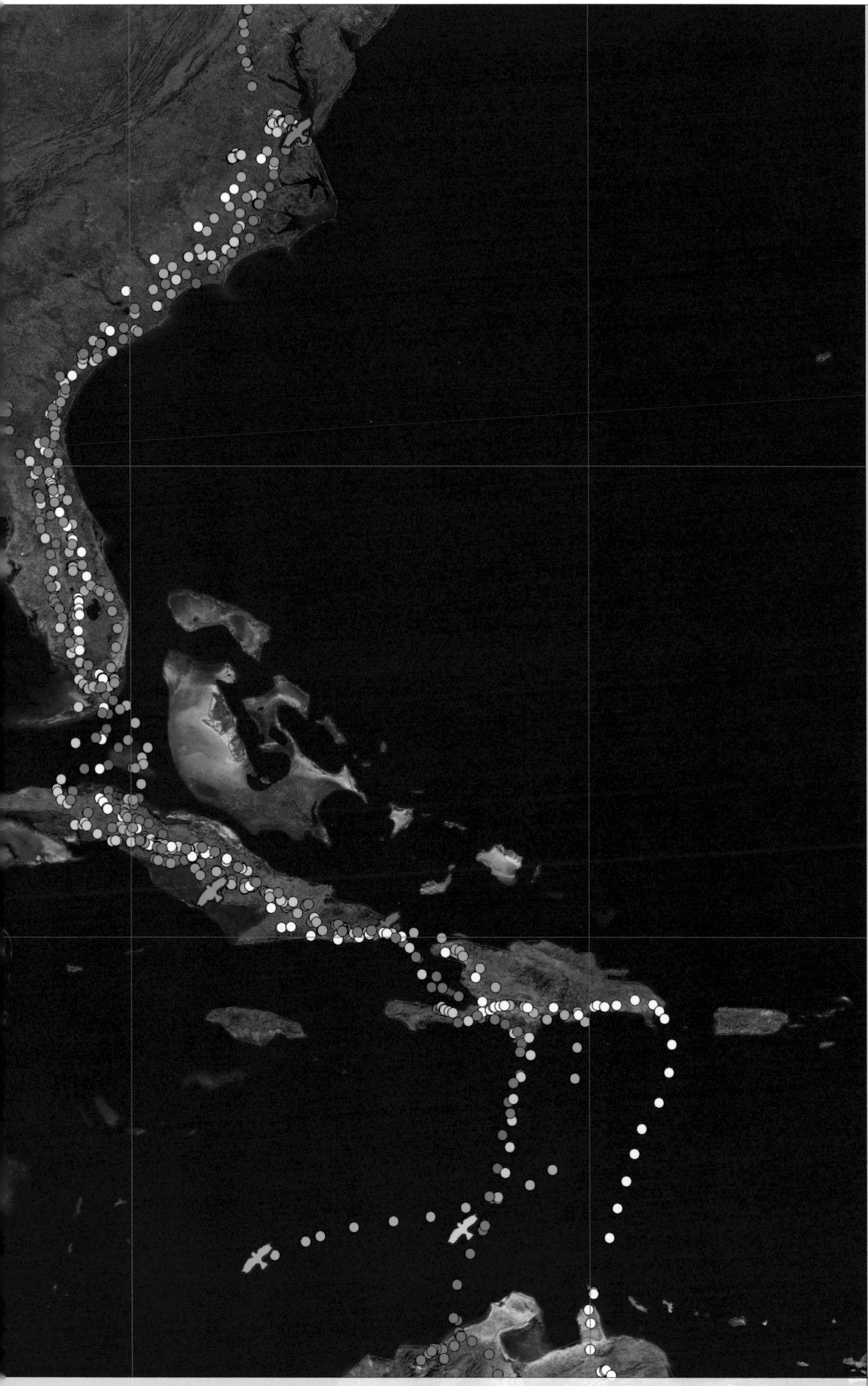

Osprey

- #46 Female
- #47 Male
- #48 Female
- #49 Female
- #50 Male
- #51 Juvenile
- #52 Male
- #54 Juvenile

The U.S. Department of Agriculture (USDA) at Langley Air Force Base (LAFB) has partnered with the GIS Team at NASA Langley Research Center (LaRC) to collect and display osprey nesting site, tracking, and reproductive data. The Legacy Osprey Relocation Project is an effort to understand the risks of osprey–aircraft collisions. Ten osprey were originally captured, outfitted with GPS transmitters, and released at selected nests surrounding Langley AFB. Different stages of the tracking include breeding, migration, and wintering. Collected data is stored in a GIS for analysis and comparison of osprey flight paths with Langley AFB flight paths and activities. Data will be collected for three years to capture local and migratory flight patterns. Eventually, the osprey data will be referenced to aircraft flight paths to develop GIS-based strike-risk models.

Courtesy of NASA Langley Research Center, GIS team.

EuroGeoSurveys—The Association of Geological Surveys of the European Union

Brussels, Belgium
By R. Salminen (chief-editor) et al.

Contact

Alecos Demetriades
ademetriades@igme.gr

Software

ArcView 3.2, Alkemia Smooth (Geological Survey of Finland)

Printer

HP Color LaserJet 8550N

Data Source(s)

Geochemical Atlas of Europe

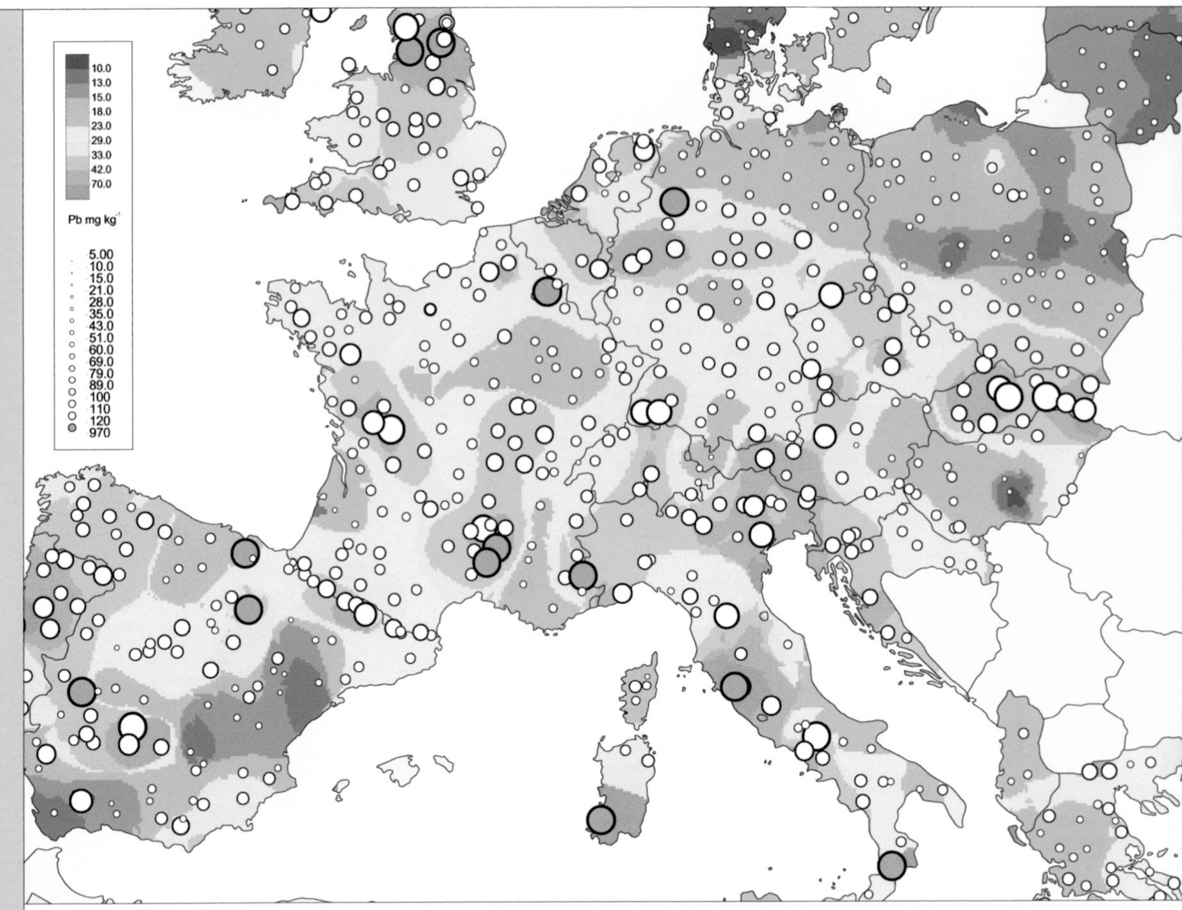

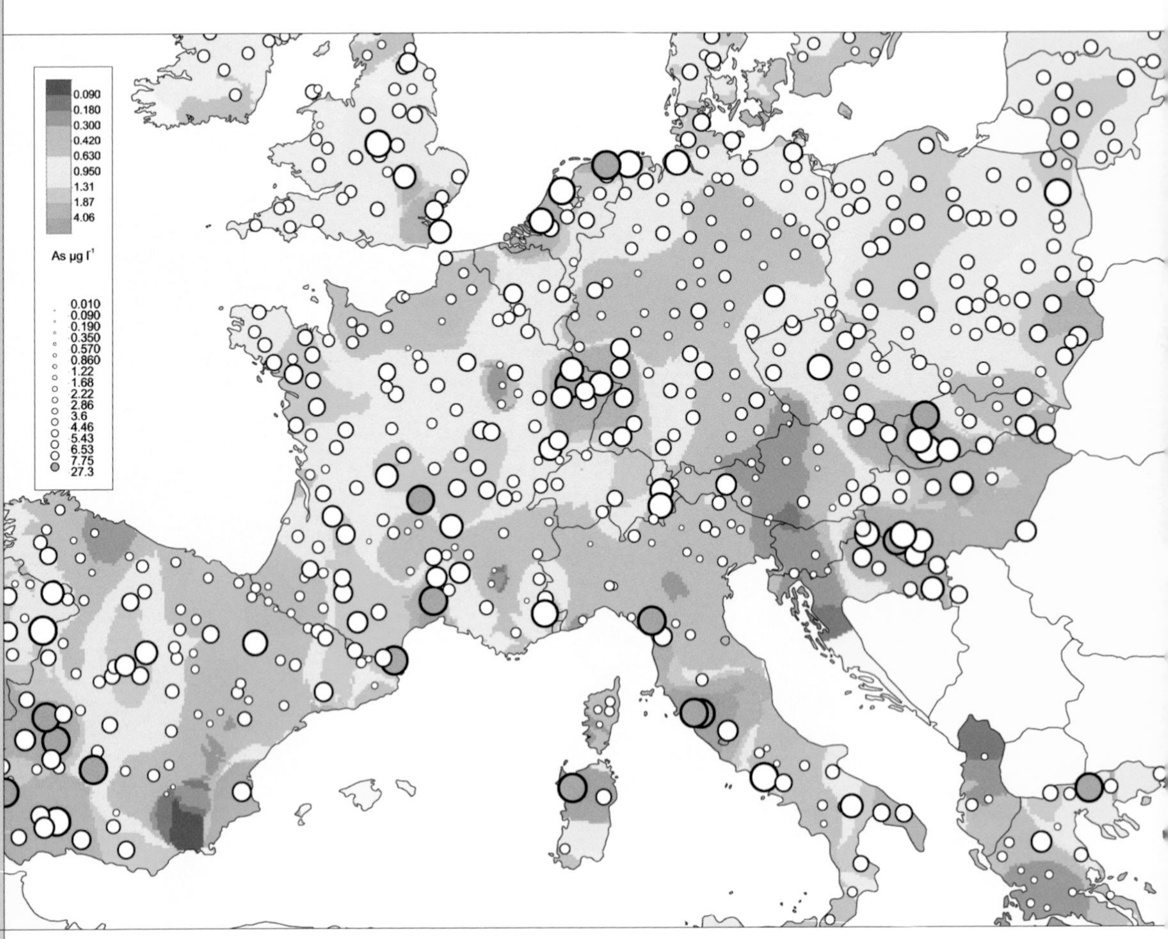

The Geochemical Atlas of Europe is the contribution of the Association of Geological Surveys of the European Union (EuroGeoSurveys) to the IUGS/IAGC Global Geochemical Baselines project. The European geochemical baseline survey covers 26 countries and provides invaluable information about the natural and human-induced concentrations of chemical elements in different sample media of the near-surface environment (topsoil, subsoil, humus, stream sediment, stream water, and floodplain sediment). This is the first multinational project performed with harmonized sampling, sample preparation, and analytical methodology, producing high-quality compatible datasets across political borders. The first phase of the project was completed and the results published in a two-volume set.

High levels of lead (Pb) and other toxic elements in the surface soil of Lavrion, Greece, were caused by years of intensive mining and smelting activities in the late twentieth century. High toxicity levels can be found in plants, animals, and humans. Children who exceed 10 µg per 100 ml of lead in their blood, the maximum amount set by the World Health Organization, suffer from low composite mental functions, and a comparative reduction in their development, especially with respect to the circumference of their head and chest.

In Croatia, Bosnia, Bulgaria, and Serbia, residents of villages lacking access to municipal (treated) water supplies are subject to a severe and potentially fatal kidney disease. This disease is known as Balkan Endemic Nephropathy (BEN). Developed by drinking contaminated water, it can lead to renal failure, requiring blood dialysis, and is often associated with kidney cancer. The principal aquifer in regions where BEN occurs is coal (Pliocene lignite), which contains many chemically reactive hydrocarbons. Scientists believe that water leaches the hydrocarbons from the coal, which naturally contaminates it.

Courtesy of the Geological Survey of Finland and EuroGeoSurveys.

Department of Environmental Science and Policy, University of California, Davis
Davis, California, USA
By Nathaniel E. Roth, Joshua H. Viers, Michele R. Slaton, Hugh D. Safford, and United States Forest Service, Region 5

Contact
Nathaniel E. Roth
neroth@ucdavis.edu

Software
ArcGIS Desktop 9.1, Microsoft Access

Printer
HP Designjet 800ps

Data Source(s)
USFS

The Terrestrial Ecological Unit Inventory (TEUI) is the standard process for ecosystem mapping used by the U.S. Forest Service (USFS). From 2004 to 2005, the USFS Pacific Southwest Region Ecology Program carried out a Landtype Association-scale TEUI for the Lake Tahoe Basin Management Unit. After completion of mapping, the Information Center for the Environment (ICE) at the University of California, Davis, carried out an assessment of climate, vegetation, soils, bedrock geology, and infrastructure as grouped by the mapping units.

ICE developed a prototype method for summary reporting on numerous datasets by individual mapping unit. ICE used ModelBuilder for ArcGIS Desktop 9.1 to create the required data in a personal geodatabase. Microsoft Access queries were used to carry out additional summary calculations and its reporting engine for document creation. Changes to input data are propagated throughout the reporting mechanism without requiring additional user effort, and reports are automatically updated.

Overall, results indicate that the method developed can be easily repeated for other national forests. The methodology also can be applied to other datasets or for other purposes. Improved versions already are under development to support other ecosystem mapping projects in northern California.

Courtesy of Nathaniel Roth, The Information Center for the Environment, Department of Environmental Science and Policy, University of California, Davis.

ENTRIX, Inc.
Walnut Creek, California, USA
By Lance Mobley, Stephen Peck, and Katie Ross-Smith

Contact
Douglas Brice
dbrice@entrix.com

Software
ArcGIS Desktop 9.1, ArcGlobe 9.1

Printer
HP Designjet 750

Data Source(s)
USGS; Tahoe Regional Planning Agency (TRPA); ENTRIX, Inc.; Placer County Department of Public Works

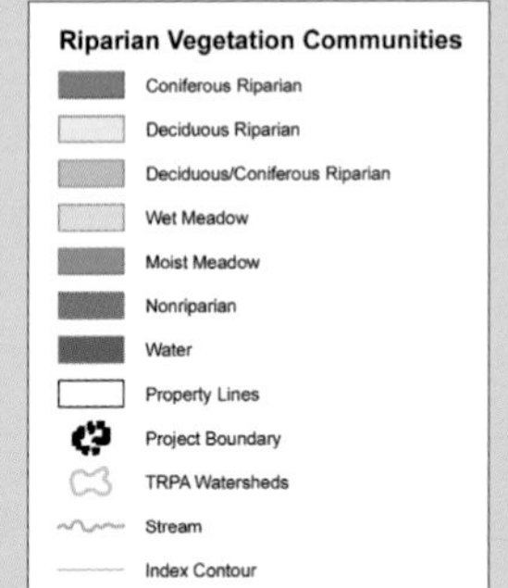

The Placer County Department of Public Works contracted Lumos and Associates, Inc., and ENTRIX, Inc., to perform a watershed analysis for the Homewood Erosion Control Project. Included in the analysis were the completion of an Existing Conditions Analysis, a Stream Environmental Zone Assessment, preliminary hydrologic and hydraulic analyses, and the development of restoration alternatives for the project area.

Located on the western shore of Lake Tahoe, Homewood comprises 625 acres of residential and commercial area adjacent to Lake Tahoe, with a 6,000-acre contributing watershed above it. Quail Lake and Madden, Ellis, and McKinney creeks comprise the major watersheds in this area, as defined by the Tahoe Regional Planning Agency.

GIS played an essential role in performing data collection and hydrologic analysis, and in identifying sensitive plant and animal areas. In particular, GIS was utilized in the following areas of analysis: vegetation mapping, stream classification, bank stability mapping, riparian vegetation assessment, watershed modeling, cultural resource site and survey mapping, and wildlife resources identification and mapping.

Alisé Géomatique
St-Jean-de-Vedas, France
By Hélène Durand

Contact
Hélène Durand
helene.durand@wanadoo.fr

Software
ArcGIS Desktop 9.2, ArcPad 7

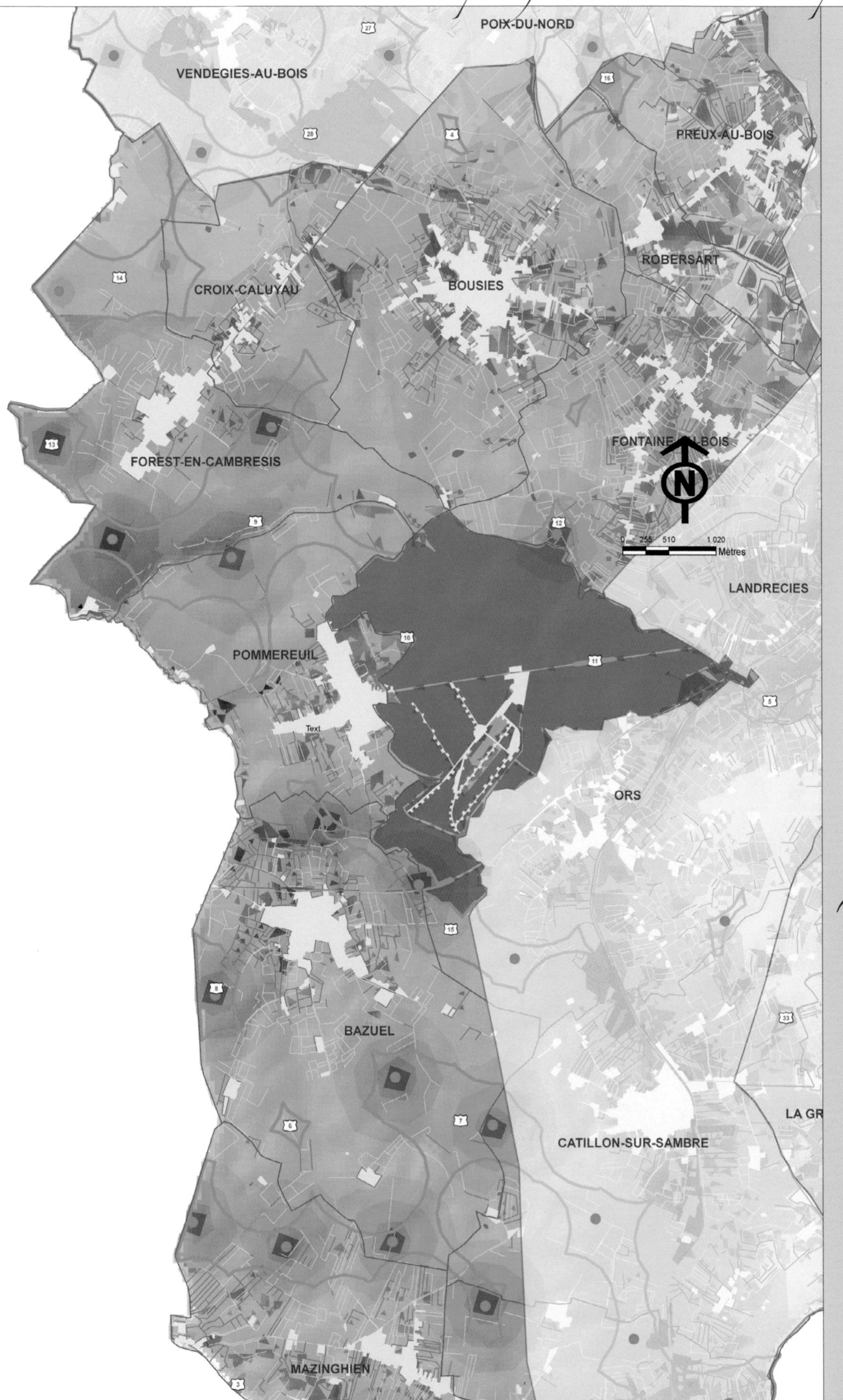

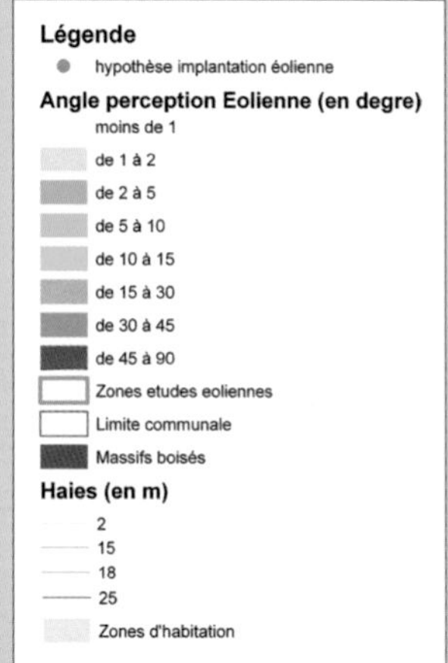

The Avesnois National Park has undertaken a wind farm development project. The park has decided to develop a plan in order to identify and classify appropriate wind zones, help the local government make informed decisions on future wind farm projects through supporting documents such as reports and maps, and create a coherent global vision for others to follow.

The installation of wind farms is a landscaping project, the aim of which is to conserve the park's diversity and unique features. Several methods were developed to analyze the park's landscape. These methods include georeferencing landscape photographs, analyzing the sensitivity of different species, studying inhabited areas, observing road and path traffic, and identifying constraints involved in the installation of wind farm equipment.

The work will contribute to the implementation of wind farm development zones and will help in the development of wind energy resources in France. The French project title is "Volet paysager du Schéma éolien de l'Avesnois."

Landforms of the Great Basin

USGS Rocky Mountain Geographic Science Center

Denver, Colorado, USA

By Harumi Warner

Contact

Harumi Warner

hwarner@usgs.gov

Software

ArcGIS Desktop 9.1

Printer

HP Designjet 5000

Data Source(s)

USGS 30-meter NED

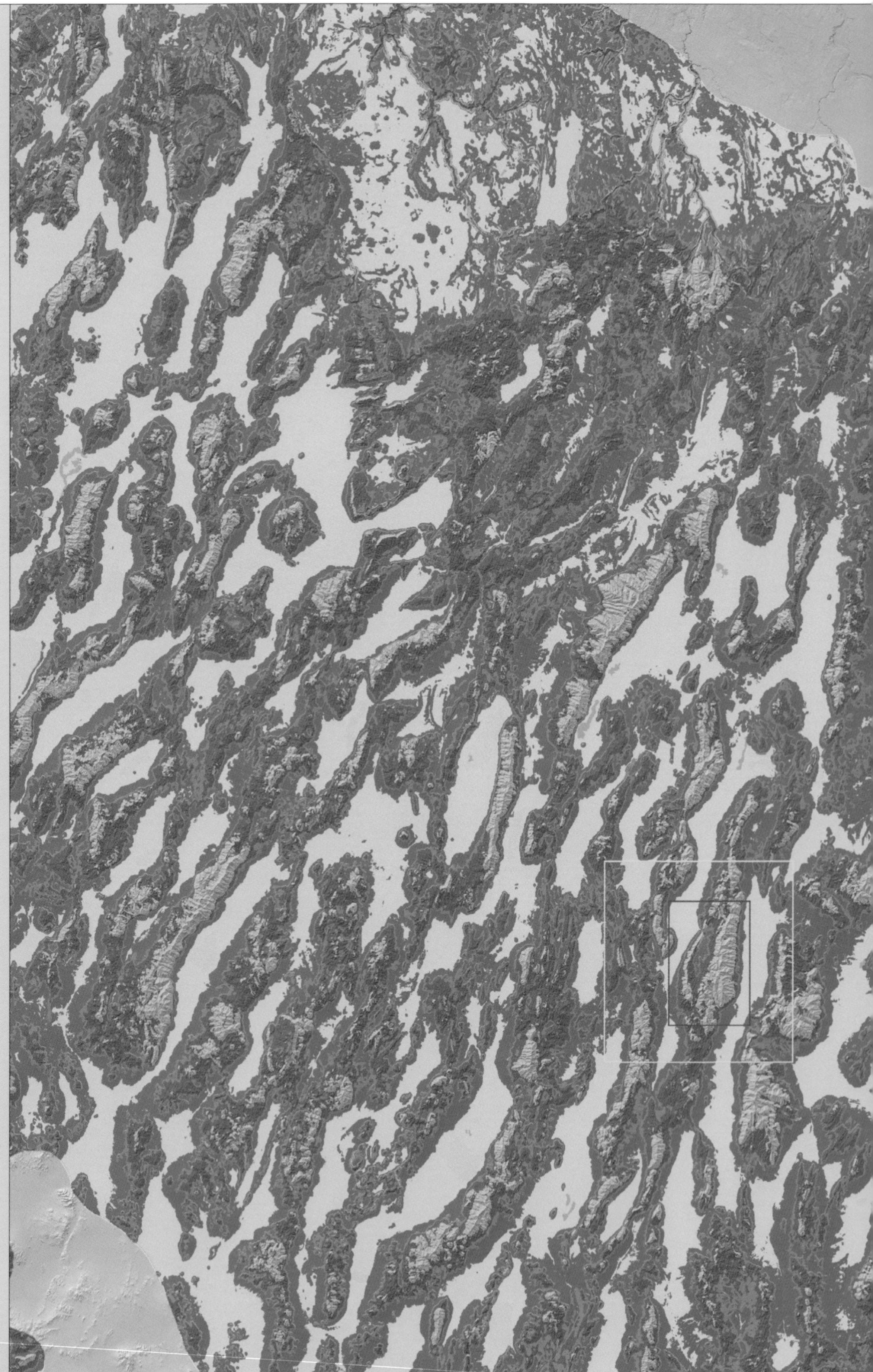

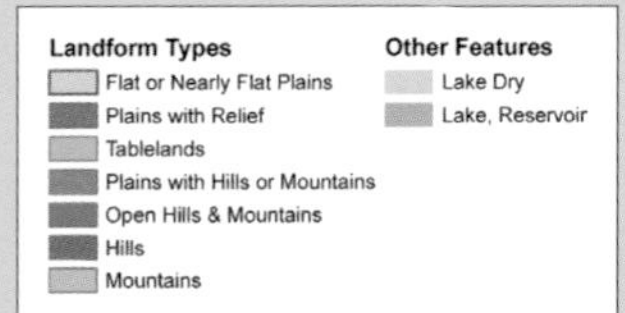

3D View

In 1964, biogeographer Edwin Hammond classified landforms in the United States by manually interpreting 1:250,000-scale topographic maps. Using a 93-square-kilometer neighborhood analysis window, he determined three input variables: slope, local relief, and profile. Unique combinations of these variables defined landform classes. By grouping the landform classes, Hammond identified five landform types: Plains, Tablelands, Plains with Hills or Mountains, Open Hills & Mountains, and Hills & Mountains. In 1991, this method was automated by Richard Dikau, Earl Brabb, and Mark Robert with GIS tools. They classified landforms in New Mexico using 200-meter digital elevation data and a 96-square-kilometer analysis window. The delineation of landforms in the Great Basin was implemented using this approach. Their algorithm was translated into a geoprocessing model in ModelBuilder for ArcGIS Desktop 9.1 and executed using the 30-meter National Elevation Dataset and a 1-square-kilometer circular analysis window. The model derived 21 landform classes in the Great Basin. Those classes were grouped into the final landform types. This map identifies seven landform types by expanding Hammond's Plains and Hills & Mountains landform types.

Courtesy of U.S. Geological Survey.

Visual Priority Routes and Use Areas

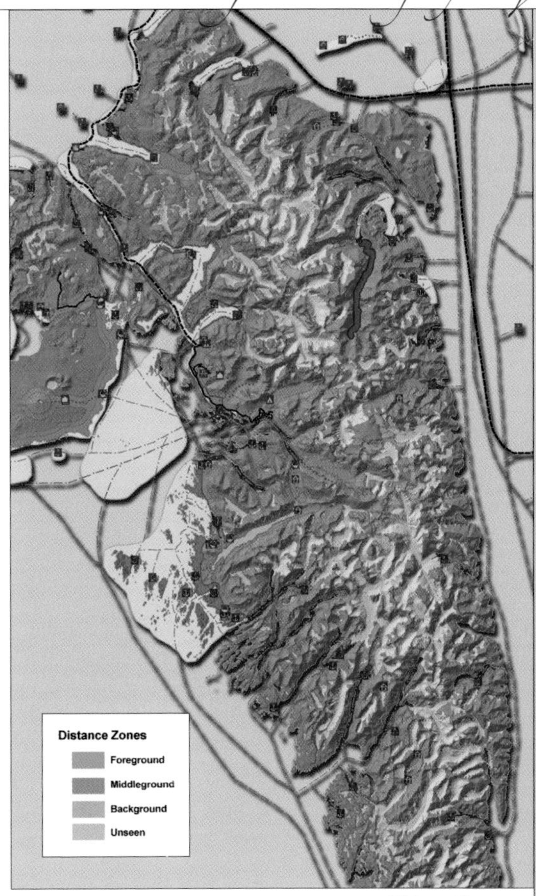

Distance Zones

Land-Use Designations

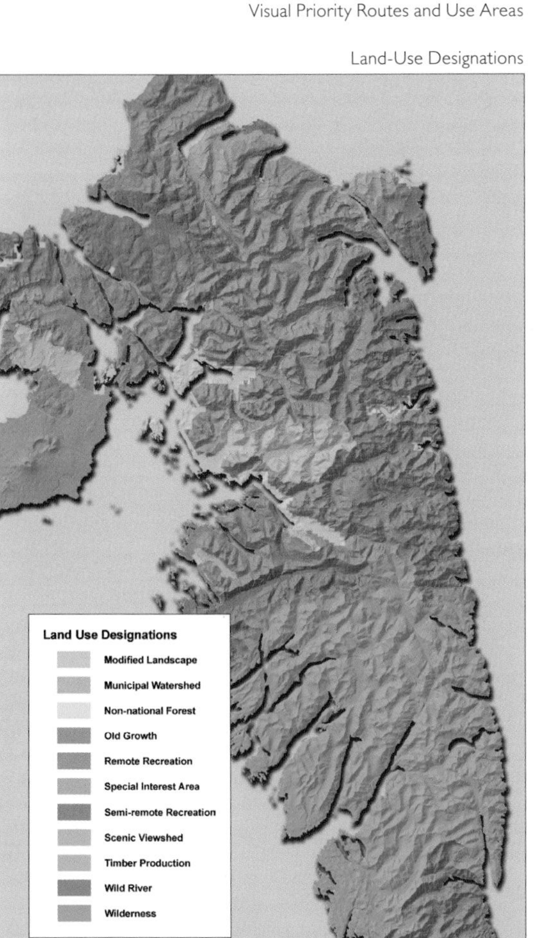

Scenic Integrity Objectives

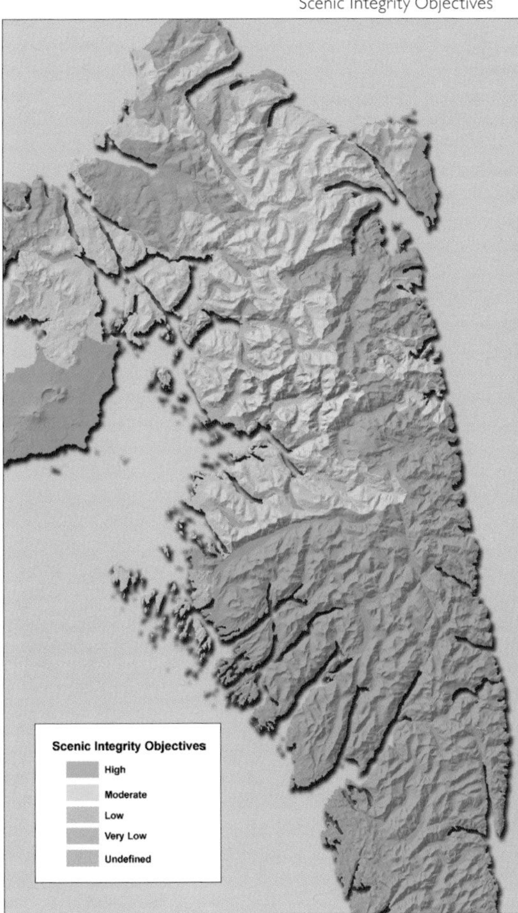

Tetra Tech EC, Inc.
Bothell, Washington, USA
By Chris Spagnuolo, Matt Kozleski, Peter Omdal, Dom Monaco, and Eric Ouderkirk

Contact
Andy Bury
andy.bury@tteci.com

Software
ArcInfo 9.1, Adobe Photoshop Elements, Adobe Illustrator

Printer
HP Designjet 5500

Data Source(s)
USGS 30-meter DEM, proprietary Tongass National Forest data

The United States Department of Agriculture (USDA) Forest Service uses the Scenery Management System (SMS) as the framework for integrating scenery management data into all levels of forest planning. A major component of the SMS is scenic integrity levels. Scenic integrity is used to describe an existing situation and to set the standards for management or desired future conditions (objectives). It is a measure of the degree to which a landscape is visually perceived to be "complete." The purpose of this analysis was to develop scenic integrity objectives for the Tongass National Forest, which will influence how the USDA conducts future activities with the potential to affect the scenic character of the landscape.

Courtesy of Tetra Tech EC, Inc.

Negative Neighborhood Indicators

City of Greensboro

Greensboro, North Carolina, USA
By Todd Hayes

Contact

Todd Hayes
todd.hayes@greensboro-nc.gov

Software

ArcGIS Desktop

Printer

HP Designjet 5500

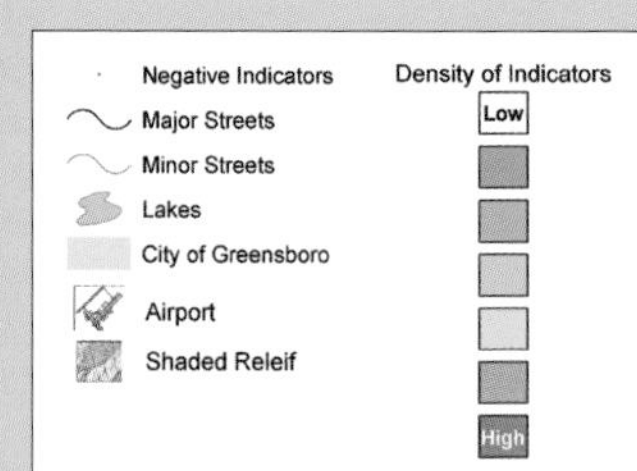

These maps depict a variety of negative neighborhood indicators throughout Greensboro, North Carolina. One of the objectives of Greensboro's comprehensive plan is to create a clean, safe, and affordable environment for all residents. Using ArcGIS software, city planners aggregated a large quantity of datasets from multiple departments and efficiently mapped these indicators. The patterns that emerge from these maps help gauge the effectiveness of city services, measure neighborhood health, and quantify existing conditions, while tracking trends over time. Using these maps as a tool, Greensboro can more effectively focus its resources and set priorities for project funding.

Courtesy of the City of Greensboro, North Carolina.

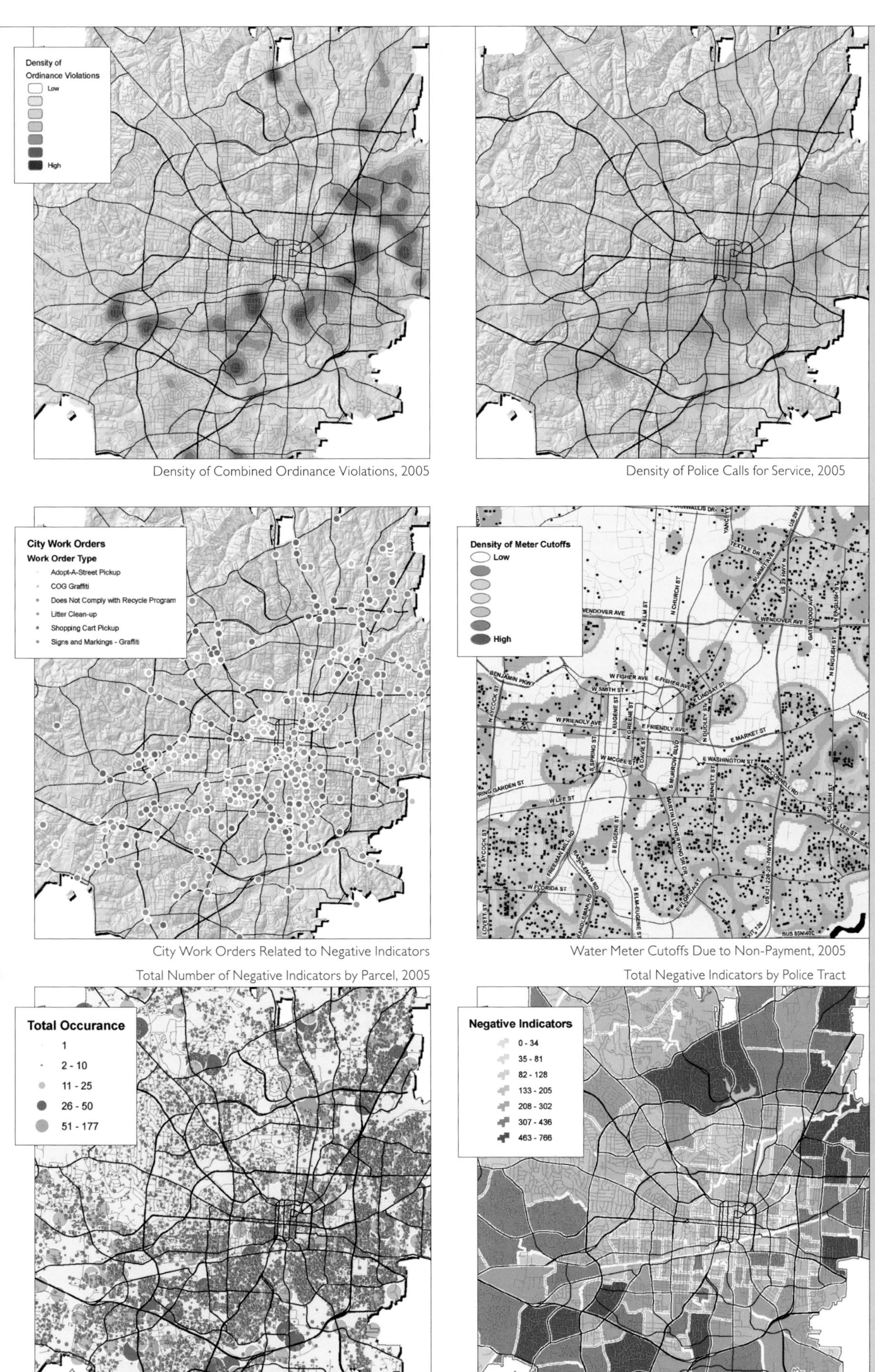

Density of Combined Ordinance Violations, 2005

Density of Police Calls for Service, 2005

City Work Orders Related to Negative Indicators

Water Meter Cutoffs Due to Non-Payment, 2005

Total Number of Negative Indicators by Parcel, 2005

Total Negative Indicators by Police Tract

Washington, D.C. Gangs

University of Maryland, College Park
Greenbelt, Maryland, USA
By Desmond Esteves and Kevin Armstrong

Contact
Desmond Esteves
desteves@wb.hidta.org

Software
ArcGIS Desktop 9.1

Printer
HP Designjet 5500

Data Source(s)
U.S. Attorney's Office, Washington, D.C., Metropolitan Police Department

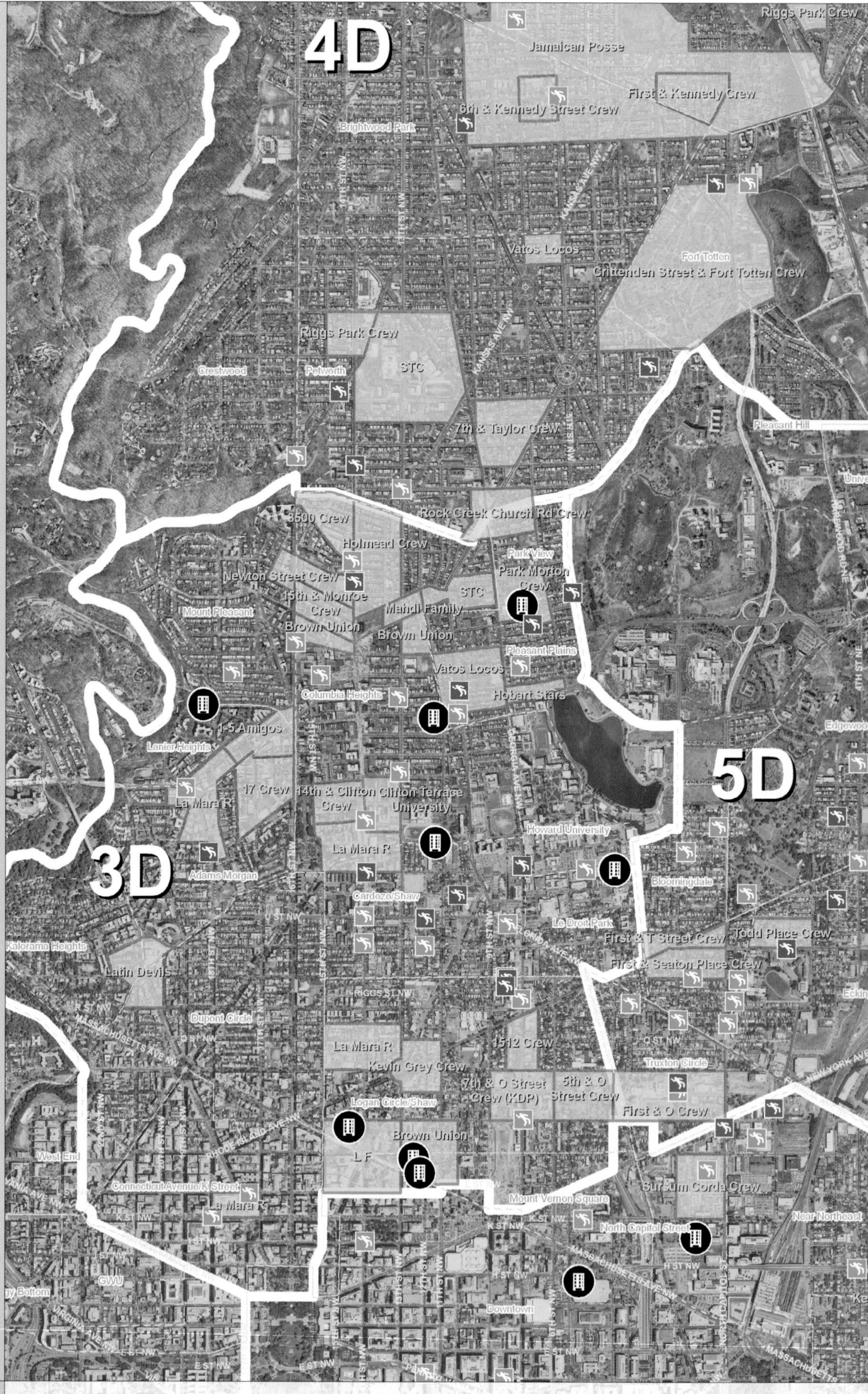

The Washington/Baltimore High Intensity Drug Trafficking Area (HIDTA) is currently working with several law enforcement agencies to address gang-related violence in Washington, D.C., and surrounding areas. The HIDTA has identified 87 gangs within the district based on interviews with police officers and intelligence gathered from suspected gang members. Many of these gangs operate in and around public housing and engage in drug trafficking. The evaluation and crime-mapping unit at the Washington/Baltimore HIDTA has mapped the locations of these gangs in relation to violent crime and public housing. For example, the map shows that about half of all homicides (135 of 272) occurred in the city's sixth and seventh districts, which contain 32 gangs. These maps are provided to various gang initiatives to aid their ongoing investigations.

Courtesy of Washington/Baltimore HIDTA.

City of Fort Collins

Fort Collins, Colorado, USA
By Marcus Bodig

Contact

Katy Carpenter
kcarpenter@fcgov.com

Software

ArcGIS 9.1, ArcGIS 3D Analyst

Printer

HP Designjet 5500

Data Source(s)

Fort Collins liquor license data, city data

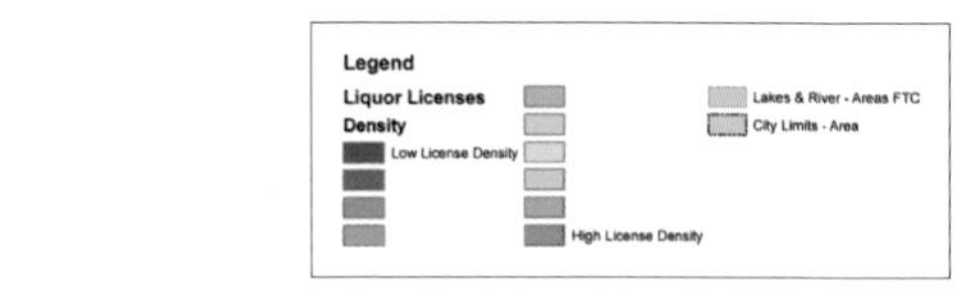

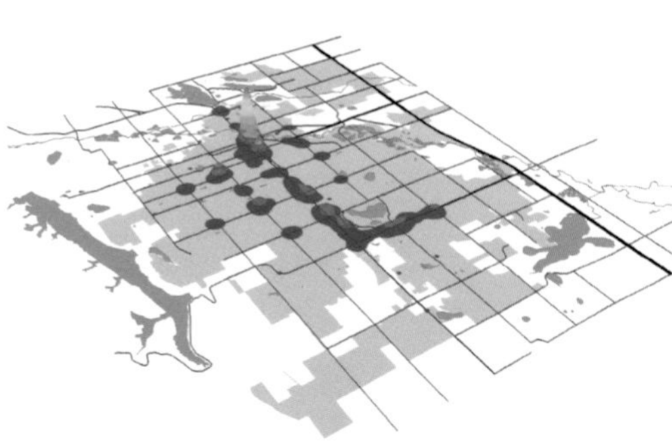

Liquor License Density

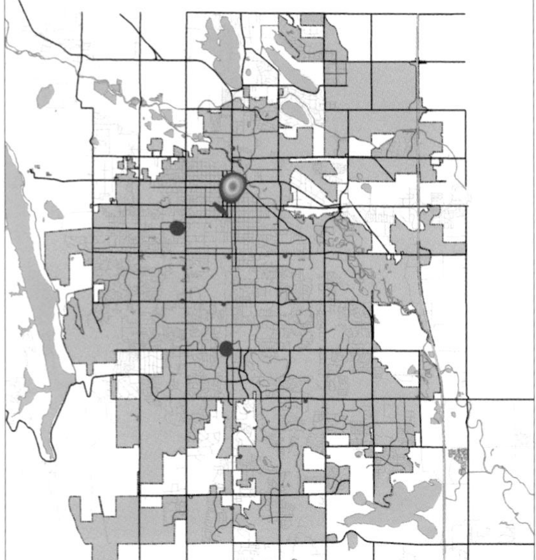

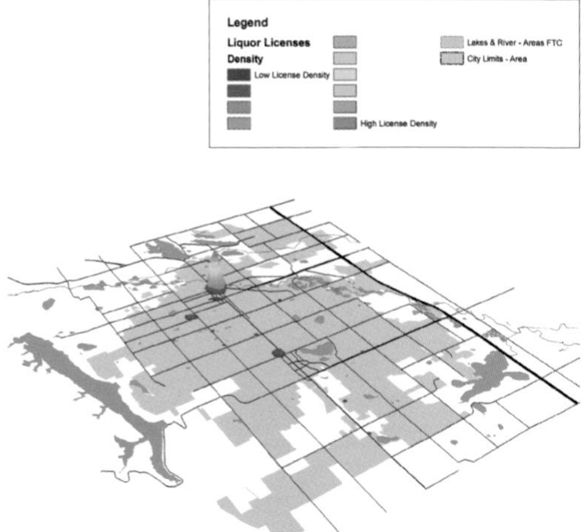

Tavern License Density

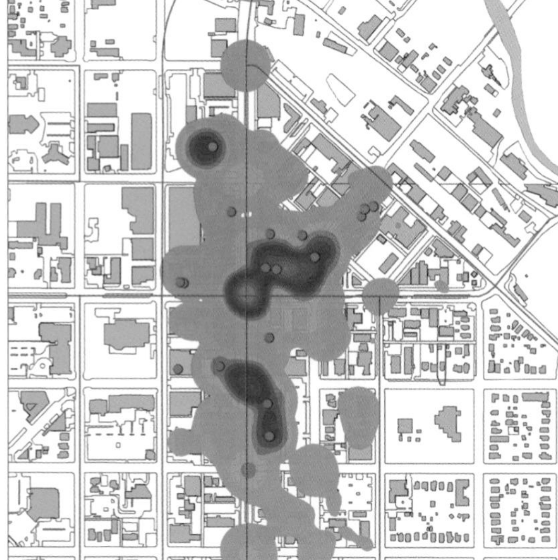

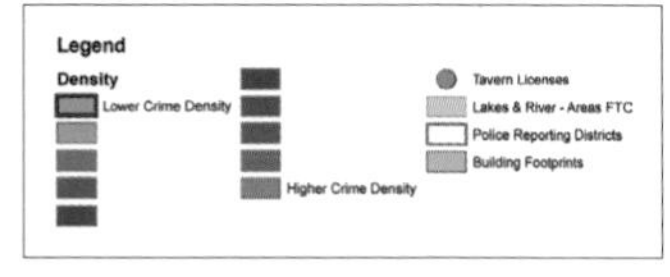

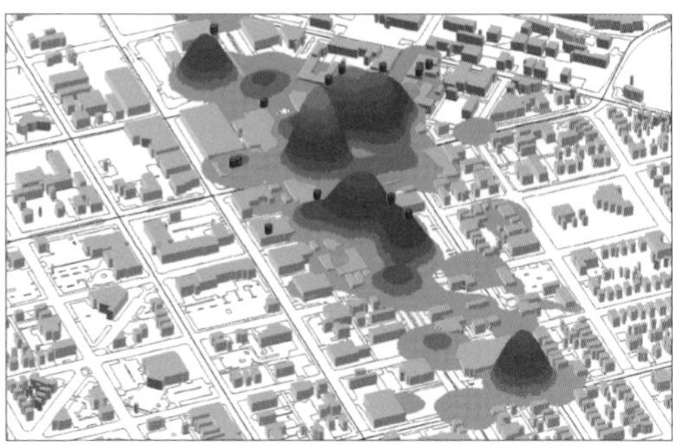

Non-Traffic Arrest Density Within Downtown Reporting Districts

In 2005, the City of Fort Collins was asked to grant a tavern liquor license to a new martini bar. The applicant proposed to occupy a 10,000-square-foot building in the historic downtown area. The downtown merchants and the police department opposed this proposal, contending there already were enough bars in the downtown area. The liquor license was denied, which prompted the proposed bar developer to take the matter to court. The municipal GIS department created these maps to help support the liquor license denial. With these maps as aids, the court upheld the downtown merchants' and police department's recommendation.

Courtesy of Marcus Bodig, City of Fort Collins, Colorado.

State of Utah, Automated Geographic Reference Center (AGRC)
Salt Lake City, Utah, USA
Mike Heagin

Contact
Matt Peters
mpeters@utah.gov

Software
ArcGIS Desktop

Printer
HP Designjet 4000ps

Data Source(s)
State Geographic Information Database (SGID)

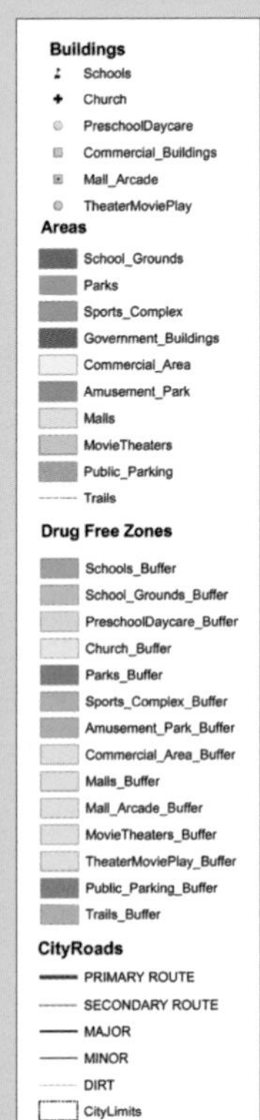

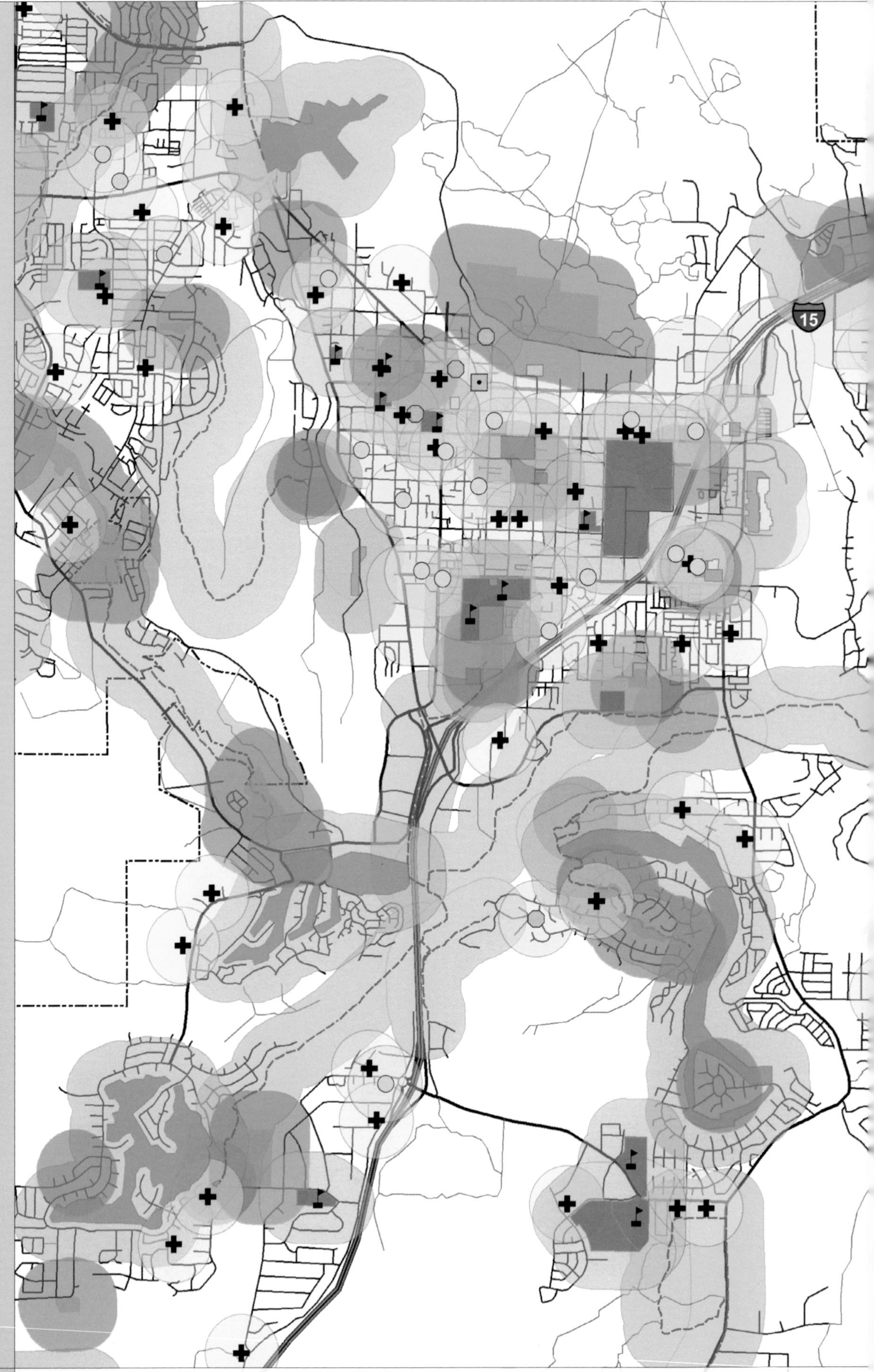

The Utah House of Representatives Law Enforcement and Criminal Justice Committee reviewed and debated the Enhanced Drug Penalty Zone Law during the 2006 legislative session. The law proposed to extend buffer zones to 1,000 feet around places where children tend to congregate. Any drug-related crimes committed in these areas would be subjected to enhanced penalties. The effectiveness of the law in achieving its original intent—to keep drug dealers away from children—was in question.

The state of Utah's Automated Geographic Reference Center (AGRC) was asked to present a visual representation of these zones in several different municipalities. The four cities that were chosen represented different types of communities in Utah, from a small rural town to a large urban city. The maps were designed to provide a visual definition of the zones, that everyone might have the same frame of reference.

Courtesy of State of Utah, Automated Geographic Reference Center.

City of Berkeley, California
Berkeley, California, USA
By Brian B. Quinn

Contact
Brian B. Quinn
bbq@ci.berkeley.ca.us

Software
ArcGIS Desktop 9.1, ArcGIS Spatial Analyst, Web NARAC

Platform
Windows Server 2003 and Linux

Printer
HP Designjet 1055cm

Data Source(s)
NARAC, USGS, City of Berkeley, BARGC

This scenario was developed as a test bed for regional disaster plumes by the City of Berkeley, California. For each of the 16 Web NARAC (National Atmospheric Release Advisory Center) consequence plumes, isoeffect contours were downloaded and vertex-densified, background value boundaries estimated, and a tension spline used in ArcGIS Spatial Analyst to create a grid model of the plume. All 16 plumes were summed to support contouring of the combined plumes. Regional National Agriculture Imagery Program (NAIP) and Landsat imagery compiled on a Bay Area Regional GIS Council (BARGC) data server was used as background imagery, and U.S. Census 2000 block population density was used to identify patterns of affected population.

Courtesy of the City of Berkeley, California.

Harris County Flood Control District

Houston, Texas, USA
By Scott Lamon

Contact

Scott Lamon
scott.lamon@hcfcd.org

Software

ArcGIS Desktop 9.1, ArcGIS Spatial Analyst, Adobe Photoshop 6

Printer

HP Designjet 5500

Data Source(s)

HCFCD HAZMAT Data, HCFCD Watershed, HCFCD Drainage Network, HCFCD 2001 LiDAR, STAR*Map Roadways

Floodplain Edition

Standard Edition

Aerial Edition

The Harris County Flood Control District's Information Services Department, through its Geographic Information Services section, utilized the ESRI developer tool, DS MapBook, to create a detailed, countywide atlas that is being used by all employees of Harris County, Texas. The map book series includes a printed and bound 512-page Standard Edition at 1:18,000 and a 1,140-page Aerial Photography Edition at 1:12,000 for departmental library use. The books cover 1,750 square miles and contain detailed hydrographical, jurisdictional, transportation, cultural, right-of-way, and parcel information for the county, which includes Houston. The map books have already proven a valuable commodity in the daily activities of the district and have become an important asset for increasing the levels of service to the communities and citizens of Harris County.

A digital version of each edition, as well as an additional version containing comprehensive floodplain information, have been developed and are available via the Internet to all employees of Harris County through a custom, browser-based application. The application is utilized daily by the Harris County Commissioners, Toll Road Authority, Public Infrastructure and Engineering, Flood Control District, Emergency Management and Homeland Security, Justice Information Management, and Public Health and Environmental Services offices, to name a few. In all, the application is being provided to approximately 11,000 Harris County employee desktops.

Courtesy of Harris County Flood Control District, Texas.

Road Network Average Speed Map

Fire Stations Quickest Response Areas Map

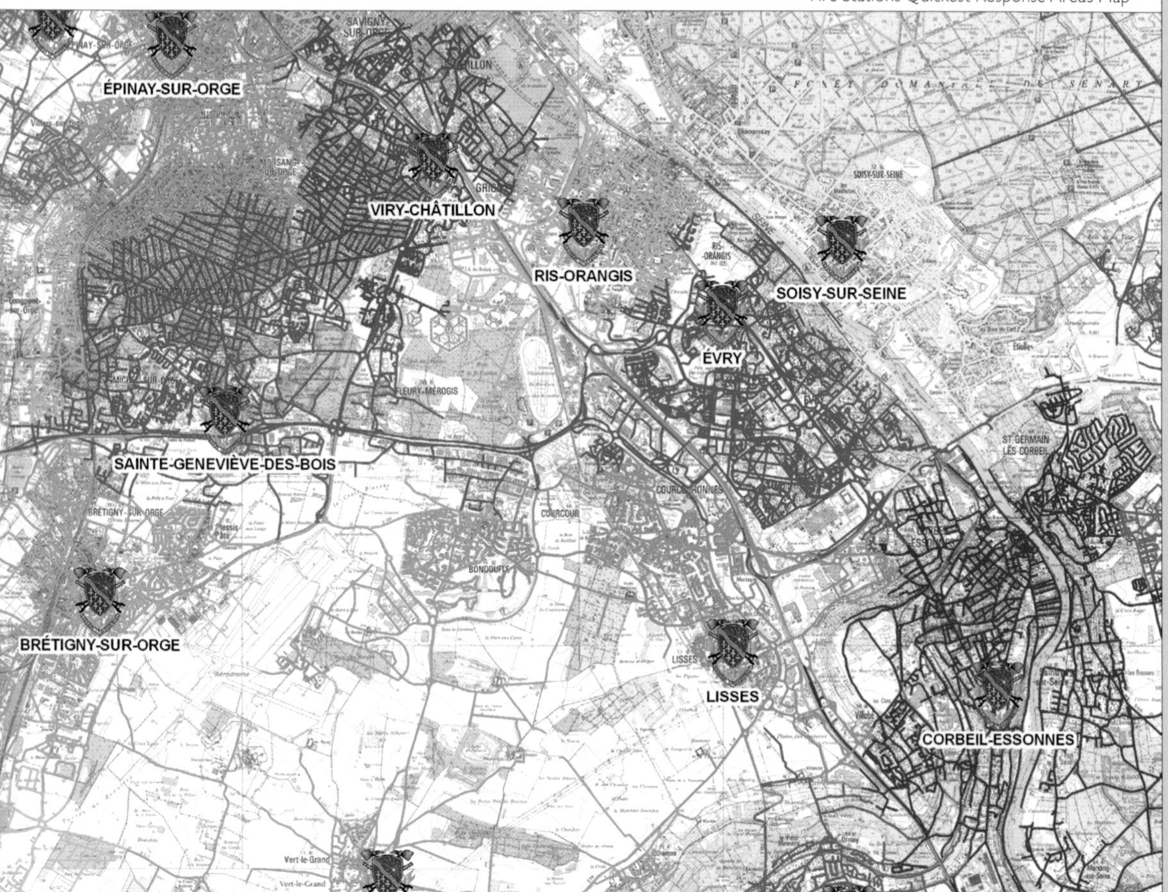

Essonne County Fire and Rescue Service (SDIS 91)

Evry, France
By Yann Kacenelen

Contact

Yann Kacenelen
ykacenelen@sdis91.fr

Software

ArcView 3.2, ArcGIS Network Analyst, ArcEditor 9.1

Printer

HP Designjet 3500cp

Data Source(s)

National Geographic Institute, NAVTEQ, SDIS 91

Firestations
10 kph
30 kph
50 kph
70 kph
90 kph

Located just south of Paris, France, Essonne County covers 710 square miles and contains a heavily industrialized northern region populated with more than a million residents. With an ever-growing number of emergency calls, the Essonne County Fire and Rescue Service (SDIS 91) has 52 fire stations responding to more than 90,000 emergency actions each year. In 2006, SDIS 91 upgraded its entire Emergency Management System, one part of it being the Computer-Aided Dispatch (CAD) system. The 15-year-old text-based software was replaced with a GIS-based system synchronized with an ArcSDE spatial database maintained by SDIS 91.

One of the main goals of using a more sophisticated CAD system was to improve the emergency response time of fire station personnel with location-based dispatching of available fire station emergency vehicles. In 2005, an ArcView Avenue script was developed to compute the three closest fire stations for every middle point of the road network's polyline in both urban and rural sample areas of the county. To improve the system and achieve more realistic results, the legal speed limit attribute in the NAVTEQ Navstreets database was dropped in favor of an average speed limit attribute. This attribute had values based on several criteria such as traffic volume, number of lanes, road category, and local ground knowledge. After extensive testing, validated calculations were carried out on 60,000 polyline road networks to produce a list of the 40 closest fire stations for each and every polyline. These lists were implemented into the CAD system as a static reference for on-the-fly calculations that take into account temporary obstacles (e.g., road closures, construction). After locating an incident, a dispatcher is able to quickly transmit emergency orders to the closest fire stations determined by the system.

Courtesy of Essone County Fire and Rescue Service (SDIS 91).

Munich Re Group, GeoRisks Research
Munich, Germany
By Lorenz Dolezalek

Contact
Lorenz Dolezalek
LDolezalek@munichre.com

Software
ArcInfo 9.1, ArcGIS Spatial Analyst

Printer
HP Designjet 1050

Data Source(s)
ESRI ArcWeb Services, SRTM, SPOT 5 (CRISP), Landsat (NASA)

A geographic information system model was developed to estimate the extent to which coastal land is likely to be inundated by tsunamis. Based on a digital elevation model (SRTM data), water flow from the coastline inland was calculated iteratively using ArcGIS 9.1 Spatial Analyst application tools. Assuming tsunami heights of 5 meters, 10 meters, and 20 meters, tsunami inundation zones were modeled for most of the world's coastlines. The hazard maps were developed for the risk management purposes of the insurance industry but could also be used by governments as a basis for land-use planning.

The satellite image shows northern Sumatra (map 1) in December 2004 when the Sumatra-Andaman earthquake created a massive tsunami that struck Indonesia and caused countless fatalities. In some areas, the tsunami peaked at over 20 meters and reached several kilometers inland, almost completely destroying the town of Banda Aceh. The modeled inundation depth for a 20-meter tsunami is indicated in shades of red. A high degree of agreement between the calculated inundation zone and the areas actually destroyed can be seen in the detailed maps (2A and 2B).

Using the model, future potential scenarios were also examined. For example, map 3 shows the modeled inundation zones of a 10-meter tsunami off the coast of Seattle in the event of a 9.1-magnitude earthquake along the Cascadian fault. The results of the scenario show that no residential areas would be affected, but the ports north and south of the city center would be hit.

Courtesy of Munich Re Group.

Map 3: Satellite image of Seattle (ArcWeb Services). The modeled inundation depths for a 10-meter tsunami are shown in shades of red.

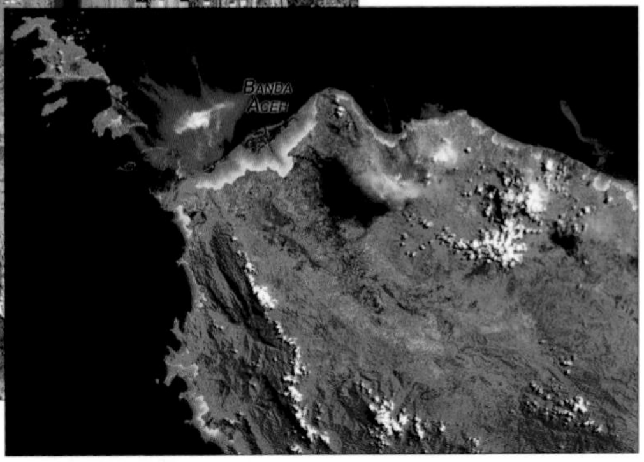

Map 1: Landsat satellite image of northern Sumatra (NASA 2000, bands 7, 4, 2). Modeled inundation depths for a 20-meter tsunami are shown in shades of red.

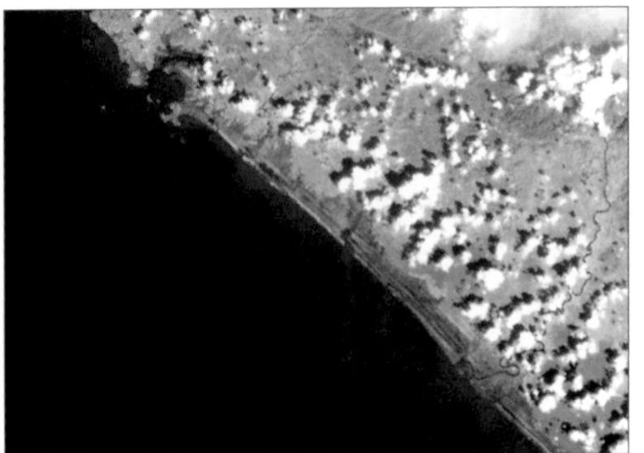

Map 2A: Detail of a SPOT 5 satellite image (CRISP 2004), recorded after the tsunami on December 29, 2004.

Map 2B: Detail of a SPOT 5 satellite image. The modeled inundation depths for a 20-meter tsunami are shown in shades of red.

An Analysis for Additional Fire Hydrant Locations

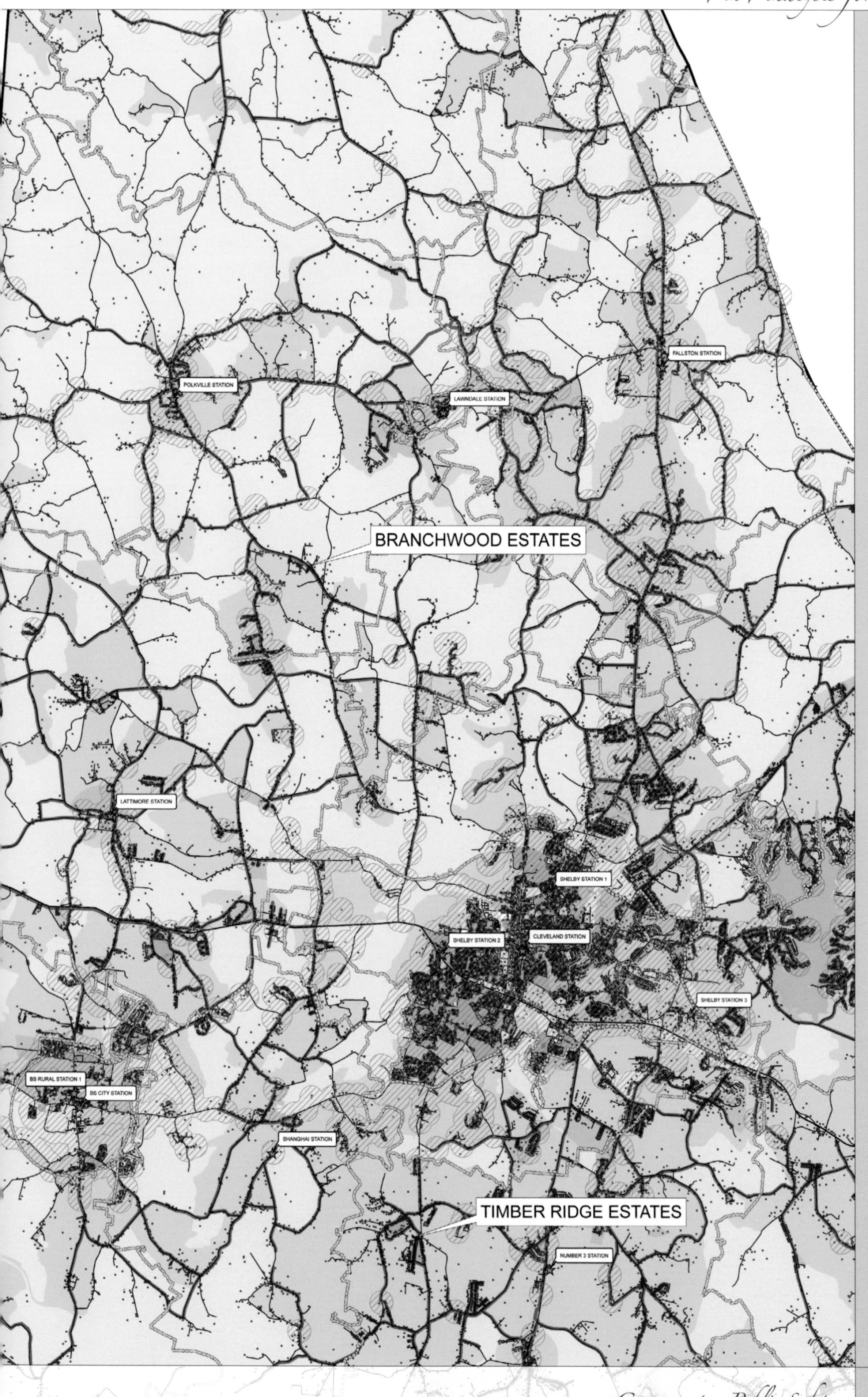

Cleveland County
Shelby, North Carolina, USA
By Alan Hamrick

Contact
Alan Hamrick
alan.hamrick@clevelandcounty.com

Software
ArcGIS Desktop 9.1

Printer
Oce TCS500

Data Source(s)
Cleveland County layers and assessors datasets, hydrants data (GPS), Cleveland County Sanitary District layers, U.S. Census blocks dataset

In an effort to broaden its fire and rescue services, Cleveland County, North Carolina, used ArcGIS 9.1 software to analyze the economic feasibility of placing additional fire hydrants in the county based on existing residences, improvement value, population density, available water resources, and proximity to fire stations. Additional fire hydrants would not only improve fire and rescue services throughout the county, but also benefit nearby property owners by securing their eligibility for lower insurance premiums. Using GIS, intersecting potential hydrant locations with an improvement valuation database can determine an annual insurance cost savings. In many cases, one year's insurance premium savings could pay for the fire hydrants.

Indicated in dark blue on the map, a six-inch waterline or greater is required to provide adequate pressure for fire hydrants. More densely populated areas are indicated on the map in darker shades of green. A point symbol represents existing structures. Fire hydrant locations were obtained by collecting GPS coordinates.

Courtesy of Cleveland County, North Carolina.

Disaster and Recovery: The Impacts of Hurricane Katrina on Gulfport, Mississippi

City of Gulfport
Gulfport, Mississippi, USA
By Tyrus Cohan

Contact
City of Gulfport, Division of GIS
gis@ci.gulfport.ms.us

Software
ArcGIS Desktop 9

Printer
HP Designjet 1055cm

Data Source(s)
City of Gulfport, FEMA, U.S. Census

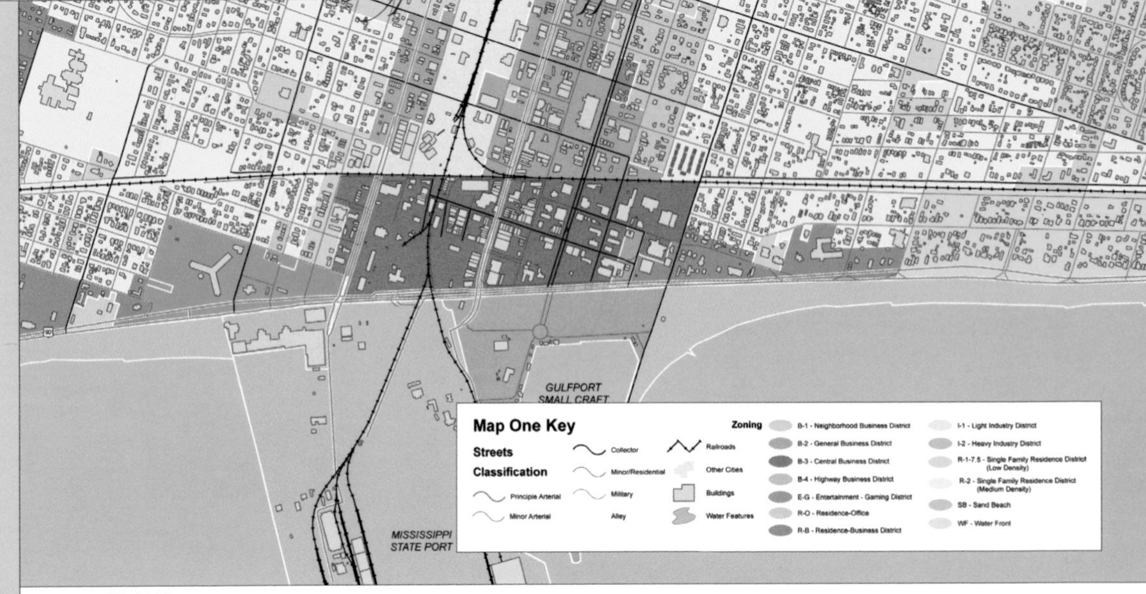

August 28, 2005

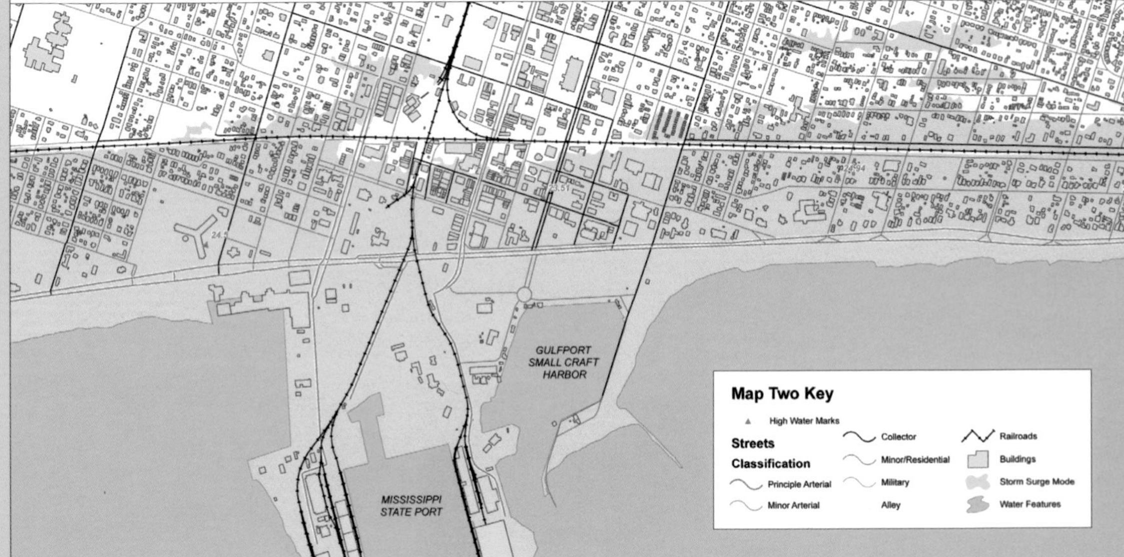

August 29, 2005—Height of Storm Surge

August 30, 2005

The Gulfport Division of GIS provides city departments and local agencies with accurate geographic information and analysis tools to increase staff efficiency and reduce redundant data collection and analysis activities among the departments. In the aftermath of Hurricane Katrina, the Division of GIS was tasked with unprecedented levels of data collection, assimilation, and dissemination. Outside agencies such as the Federal Emergency Management Agency (FEMA), the National Guard, clean-up contractors, emergency response teams, and humanitarian organizations all drew on the division for support. This map represents some of the many datasets developed as part of the recovery effort on the Mississippi Gulf Coast.

Courtesy of the City of Gulfport, Mississippi.

Riley County Customized DFIRM
Base Flood Elevations
Cross Sections
Community Boundaries
Firm Panel Boundaries
Flood Zone
Boundaries
0.2 PCT ANNUAL CHANCE FLOOD HAZARD
1 PCT ANNUAL CHANCE FLOOD HAZARD
FLOODWAY
LIMIT OF STUDY
ZONE D
Flood Zone
Areas
0.2 PCT ANNUAL CHANCE FLOOD HAZARD
A
AE
D
Streams and Shorelines
Roads and Railroads
Bridges and Levees
Section-Township-Range Lines
City of Ogden
Dry Branch
Seven Mile Creek
GREENWOOD DR
FOURTH ST

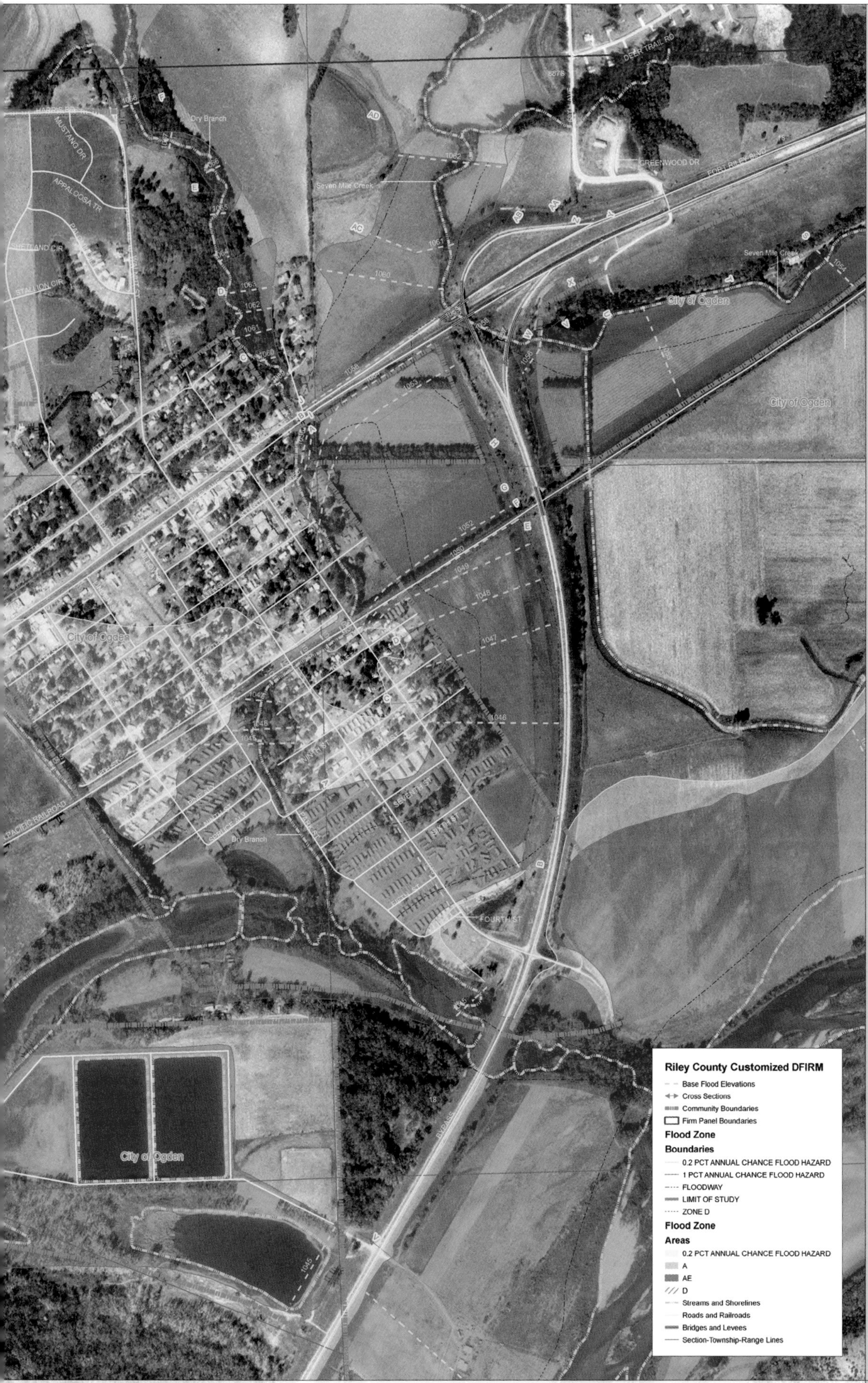

Federal Emergency Management Agency (FEMA)
Washington, D.C., USA
FEMA

Contact
Melis Mull
melis.mull@dhs.gov

Paul Rooney
paul.rooney@dhs.gov

Software
ArcGIS Desktop 9.1, Adobe Photoshop

Printer
HP Designjet 1055cm

Data Source(s)
FEMA

FORT RILEY
MILITARY
RESERVATION
ZONE X
ZONE AE
City of Ogden
200301
Riley County
Unincorporated Areas

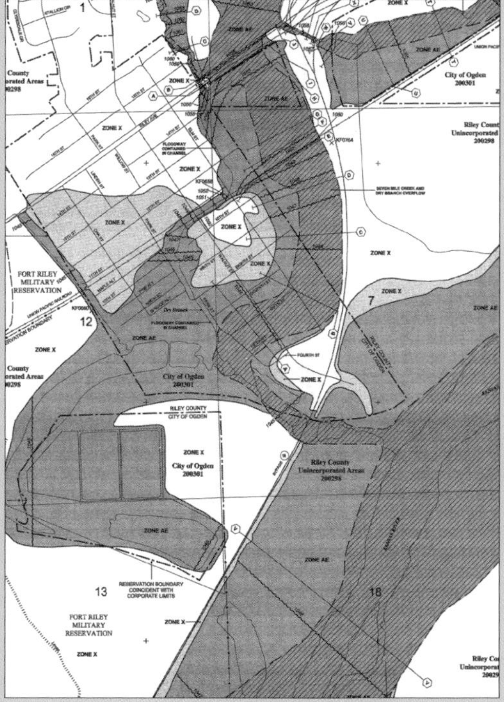

The Federal Emergency Management Agency's flood hazard maps display flood risk. This information is used to support local floodplain management, permitting the setting of flood insurance purchase requirements and insurance ratings for the national Flood Insurance Program. The Flood Map Modernization program creates updated GIS flood maps such as the paper flood map of Riley County, Kansas, shown above, which is designed for consistent, cost-effective reproduction. To the flood map on the left, a user has added an aerial photo and customized the flood data to make the map easier for county planners to use. Communities might also add zoning data, drainage basins, parcel data, evacuation routes, critical facilities, or other data to help them manage their risk.

Courtesy of Federal Emergency Management Agency.

Assessing the Vulnerability of U.S. Livestock to Agricultural Biological Terrorism

Kansas State University
Manhattan, Kansas, USA
By Thomas J. Vought Jr. and J. M. Shawn Hutchinson

Contact

Thomas J. Vought Jr.
tvought@k-state.edu

Software
ArcGIS Desktop 9.1, Adobe Illustrator 9

Printer
HP Designjet 800ps

Data Source(s)
USDA NASS 2002 Census of Agriculture,
U.S. EPA Level III Ecoregions

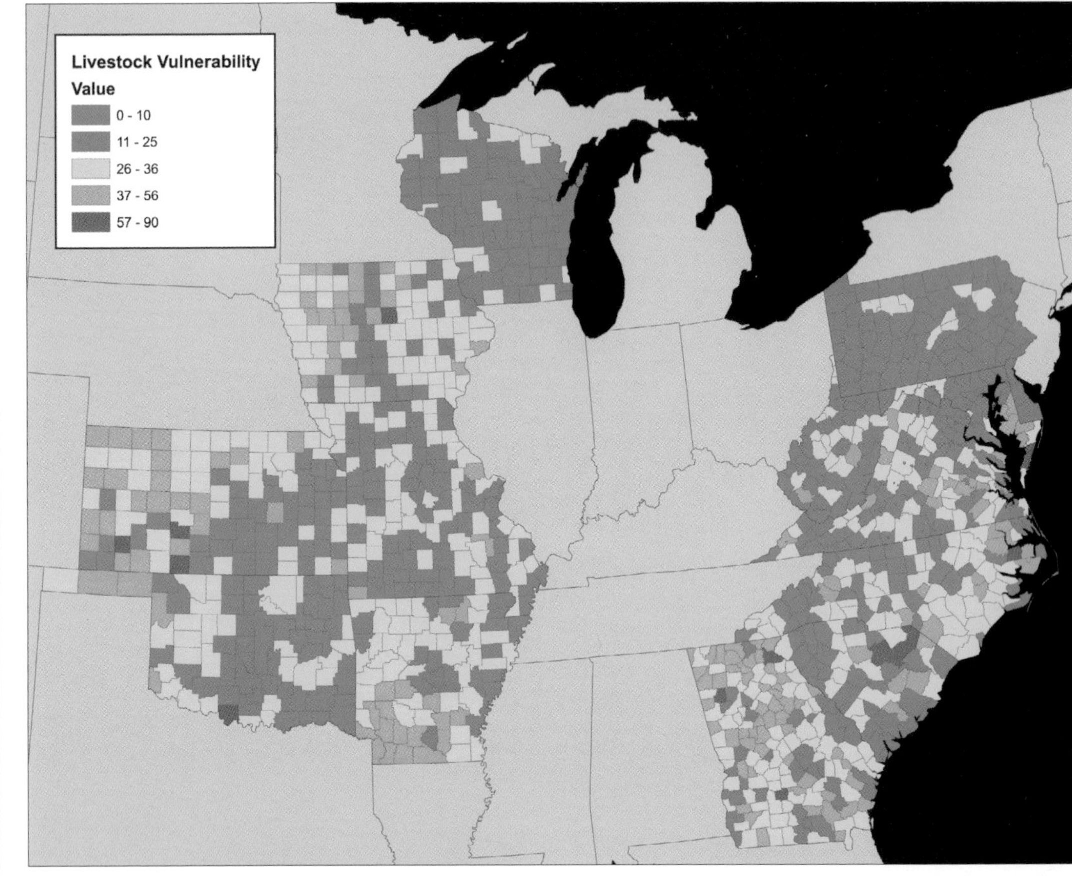

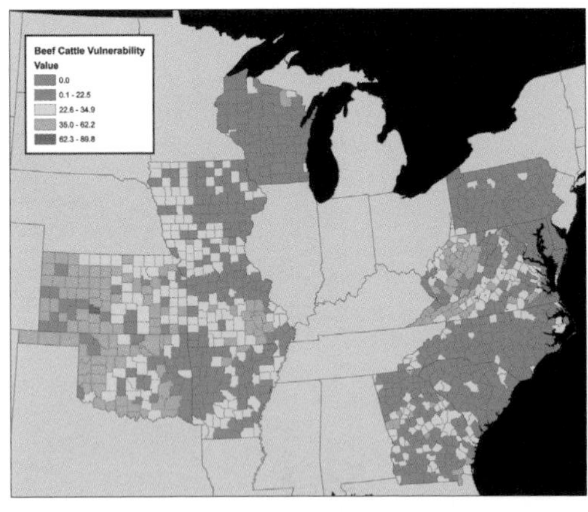

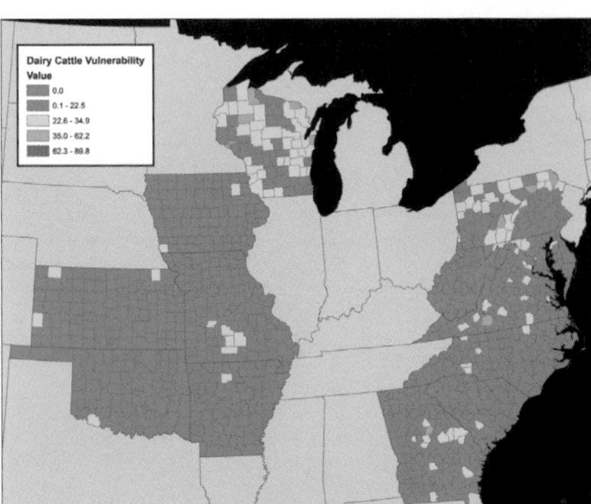

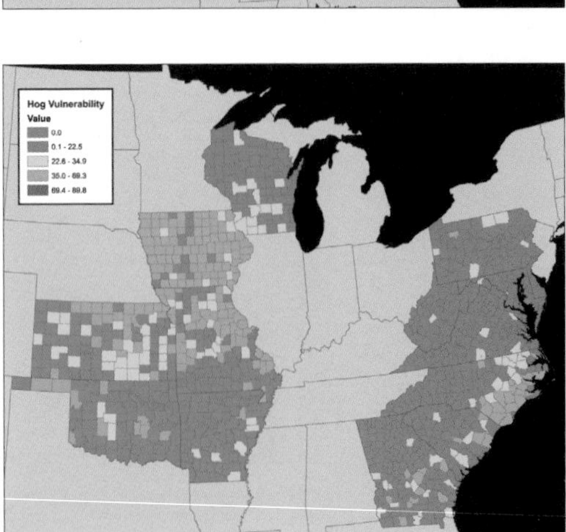

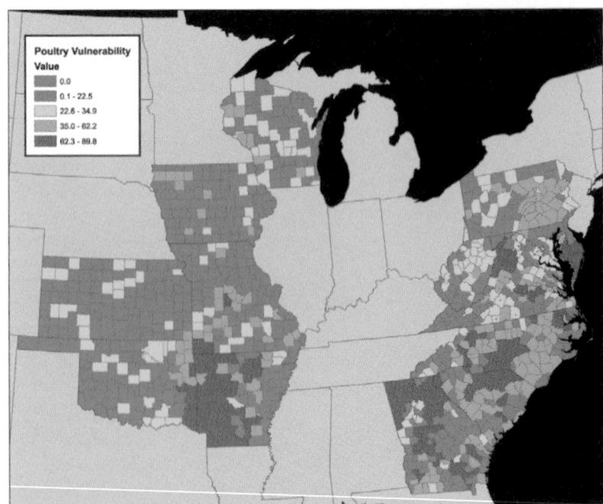

Geographers have helped identify how place, culture, and environment are critical components to better understanding the causes of terrorism and how to best respond. However, less attention has been focused on spatially explicit aspects of homeland security and the susceptibility of rural landscapes threatened by agricultural bioterrorism or naturally occurring crop and livestock diseases. This map uses data from the 2002 U.S. Census of Agriculture (USDA National Agricultural Statistics Service) to create a county-level assessment of vulnerability based upon the spatial arrangement and importance of specific agricultural commodities. Vulnerability is defined by an additive model that incorporates measures of dominance and economic importance for a specific animal type within a county, and regional diversity in livestock species.

Results show surprising regional and state-level patterns of county vulnerability to livestock disease outbreaks that would not be apparent if such an assessment were based upon statistical measures of production alone. Such an analysis should be helpful in preparedness planning and calling attention to the security concerns for our less populated places.

Courtesy of Department of Geography, Kansas State University.

Pacific Disaster Center
Kihei, Hawaii, USA
By Todd Bosse, Chris Chiesa, Pam Cowher, and Rich Nezelek

Contact
Todd Bosse, Pam Cowher
tbosse@pdc.org, pcowher@pdc.org

Software
ArcSDE 9.1, ArcIMS 9.1, ArcInfo 9.1, Adobe Photoshop CS2

Platform
Sun Solaris (server-side), Windows XP (client-side)

Printer
HP CP2800

Population Density, 2001

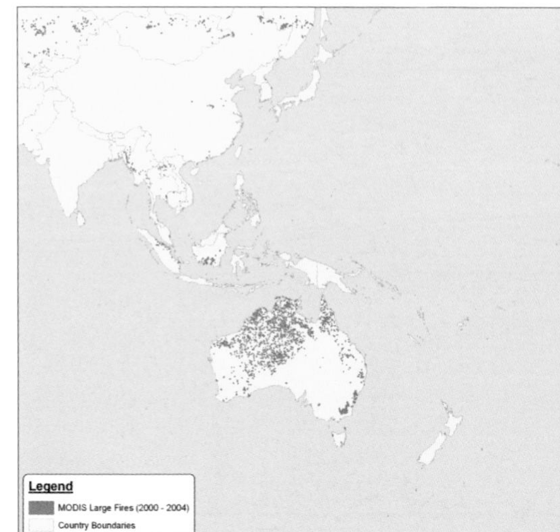

Historical Large Wildfires

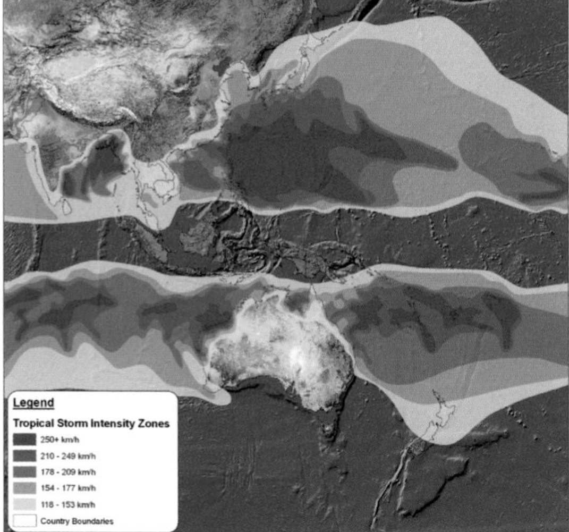

Tropical Storm Intensity Zones

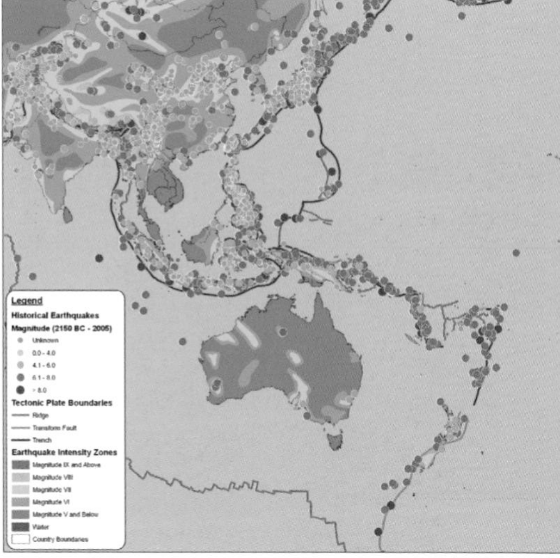

Historical Earthquake Activity

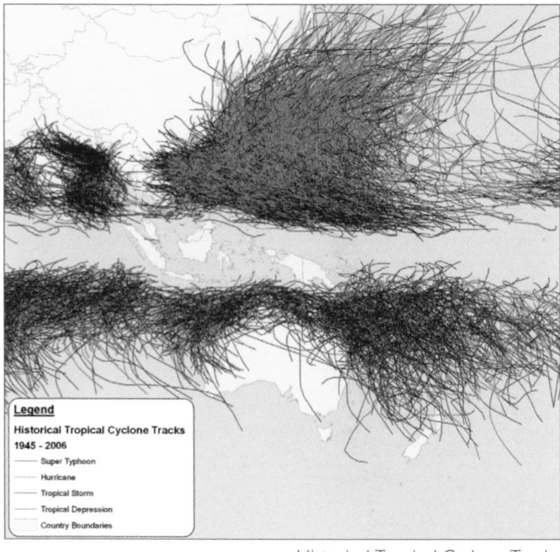

Historical Tropical Cyclone Tracks

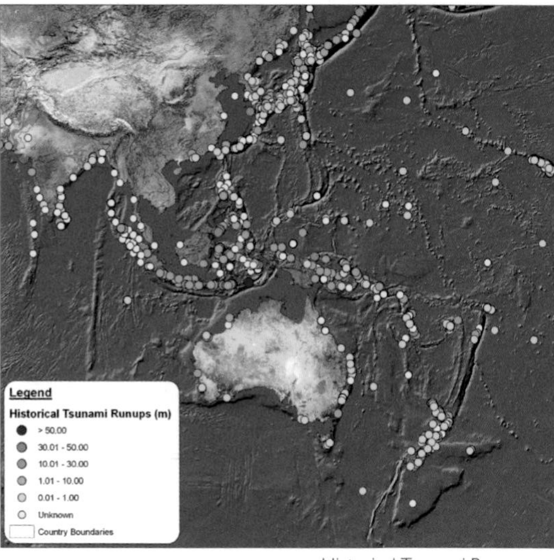

Historical Tsunami Runups

Recognizing that natural disasters are predominantly local issues, often with national, regional, or global impacts, the Pacific Disaster Center (PDC) has developed the Asia Pacific Natural Hazards and Vulnerabilities Atlas. This Hazards Atlas is an interactive, geospatial mapping tool and database for natural hazards and vulnerabilities information within the Asia Pacific and Indian Ocean regions. A principal objective of the Hazards Atlas is to provide decision makers with greater awareness of the risks of natural hazards in their area of concern and a venue for exploring regional- and national-level issues related to risk and vulnerability, as well as for assessing impacts of natural hazard events.

The atlas utilizes geographic information systems technologies to integrate and display locations, magnitudes, and dates of hazard events; delineate regional natural hazard phenomena; and depict historical trends. It also gives users access to baseline geographic and infrastructure data layers, as well as historical and near-real-time data on natural hazard events including earthquakes, wildfires, tsunamis, volcanoes, and tropical storms. The Asia Pacific Natural Hazards and Vulnerabilities Atlas combines data compiled from various sources including ESRI, NOAA, USGS, NGA, National Weather Service, Naval Pacific Meteorology and Oceanography Center's Joint Typhoon Warning Center, Oak Ridge National Laboratory, and University of Hawaii. The online atlas is available at http://atlas.pdc.org.

Courtesy of Pacific Disaster Center.

Underground Hazardous Material Sites in the Floodplains of White Oak Bayou

Harris County Flood Control District
Houston, Texas, USA
By Scott Lamon

Contact

Scott Lamon
scott.lamon@hcfcd.org

Software

ArcGIS 9.1, ArcGIS Spatial Analyst, Adobe Photoshop 6

Printer

HP Designjet 5500

Data Source(s)

2004 TSARP Preliminary Floodplains, TSARP LiDAR/2-Foot Contour Data, 2001 HCFCD LiDAR, HCFCD Drainage Network, HCAD Parcel Data, STAR*Map Roadways, HCFCD HAZMAT Data, HCFCD Watershed

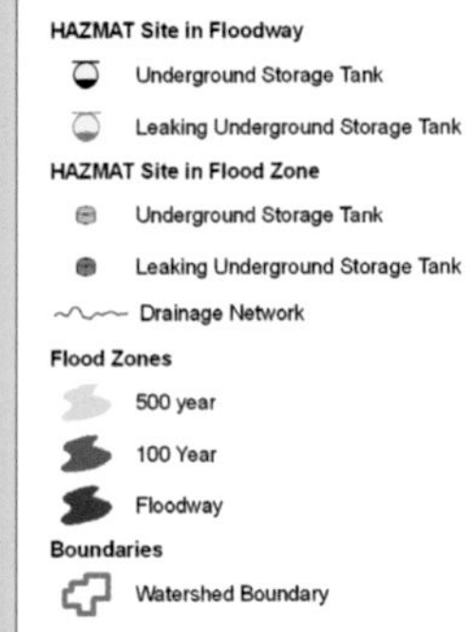

To reduce the environmental damage done by flooding within the county, the Harris County Flood Control District Environmental Services Department has developed a watershed environmental baseline dataset to be used with numerous other GIS datasets. Roughly 20 percent of the 32,000 sites identified in Harris County are underground storage tanks containing a variety of potentially harmful substances. More than 2,000 have been categorized as leaking tanks. Flooding in the areas around these tanks has the potential to cause significant damage to the downstream environment and poses a health risk to residents.

This map displays underground storage tanks (USTs) and leaking underground storage tanks (LUSTs) in the floodplains of one of the more densely populated watersheds in Harris County: White Oak Bayou Watershed. Knowing where these sites are in relation to the floodplain gives the HCFCD environmental department the ability to quickly and effectively identify existing or potential conditions that might influence the design or the construction of a flood management project.

Courtesy of Harris County Flood Control District, Texas.

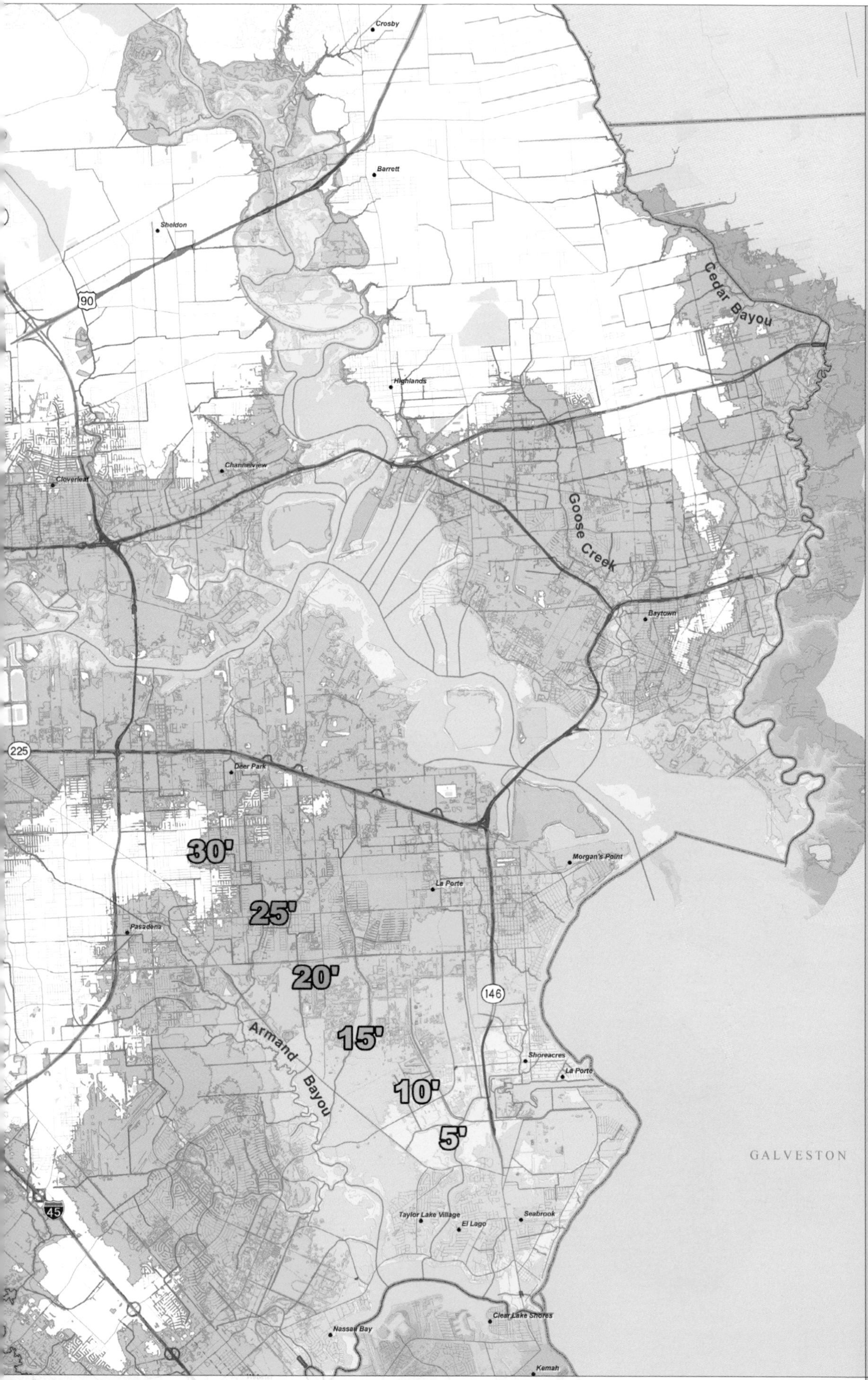

Legend

- 5' Elevation
- 10' Elevation
- 15' Elevation
- 20' Elevation
- 25' Elevation
- 30' Elevation

Harris County Flood Control District
Houston, Texas, USA
By Michael Walters

Contact
Michael Walters
michael.walters@hcfcd.org

Software
ArcGIS 9.1, ArcGIS Spatial Analyst, Adobe Photoshop 6

Printer
HP Designjet 5500

Data Source(s)
TSARP LiDAR/2-Foot Contour Data, 2001 HCFCD LiDAR, HCFCD Drainage Network, STAR*Map Roadways, Harris County Water Bodies

The Harris County Flood Control District has created a highly detailed map of storm surge inundation zones that would result from a hurricane landfall on western Galveston Island, just 50 miles southeast of downtown Houston, Texas. Utilizing elevation data derived from LiDAR, the district's GIS staff created multiple polygon features based on two-foot contours that correspond to storm surge elevations for Category 1 through Category 5 hurricanes. The resulting information can be used as an overlay of appraisal district parcel data, enabling identification of individual residences and businesses that may be in harm's way. This map, combined with the Emergency Evacuation Routes map from the Texas Department of Transportation, can provide Texas Gulf Coast citizens with valuable information that may save lives during future storms.

Courtesy of Harris County Flood Control District, Texas.

Ground Elevations Compared to Static Base Flood Elevations for the City of Seabrook

Harris County Flood Control District
Houston, Texas, USA
By Jason Cowart, Jessica Wade

Contact
Jason Cowart
jason.cowart@hcfcd.org

Software
ArcGIS 9.1, ArcGIS Spatial Analyst, Adobe Photoshop 6

Printer
HP Designjet 5500

Data Source(s)
2004 TSARP Preliminary Floodplains, 2001 HCFCD LiDAR, HCFCD Drainage Network, HCAD Parcel Data, STAR*Map Roadways

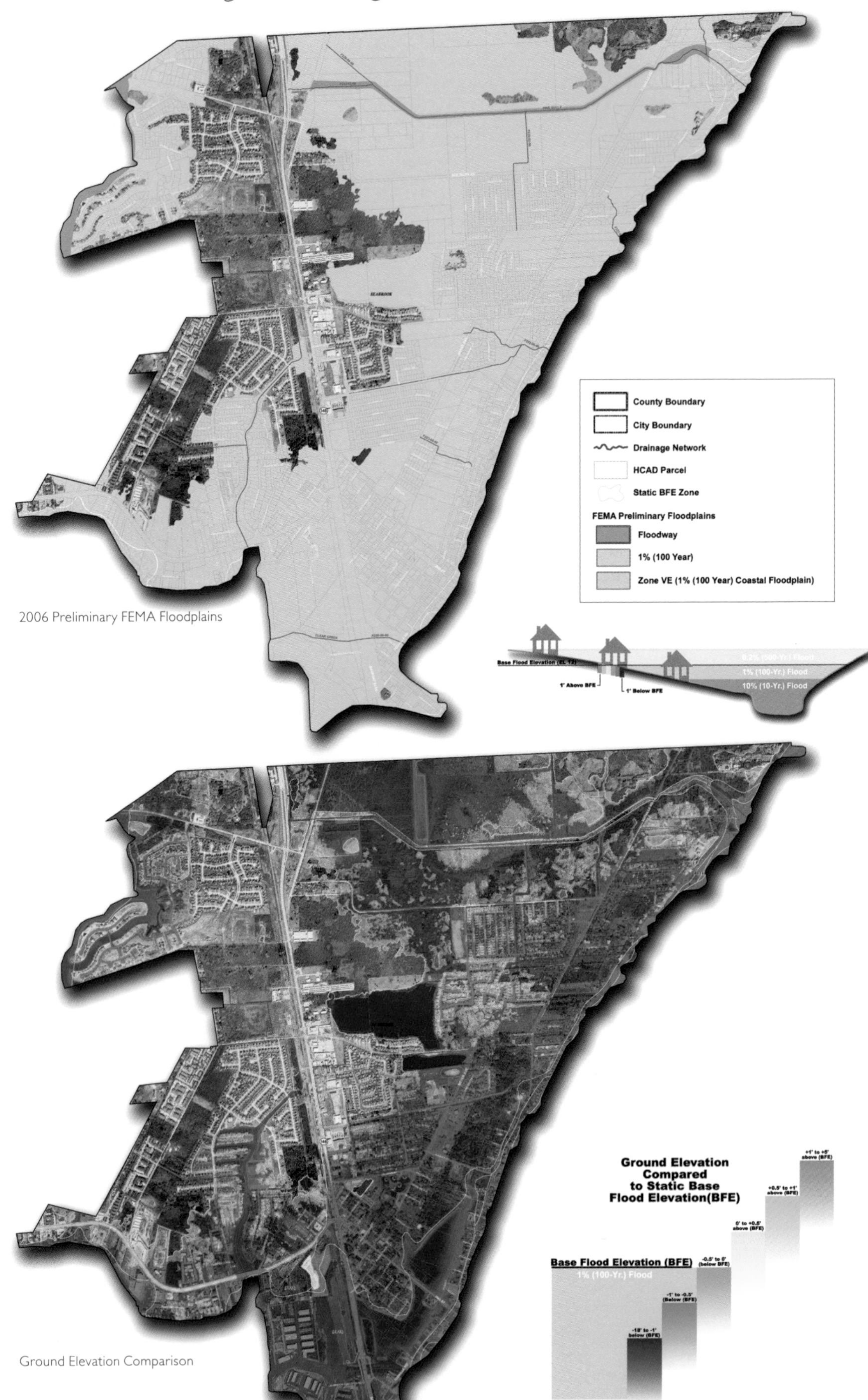

2006 Preliminary FEMA Floodplains

Ground Elevation Comparison

The City of Seabrook requested that the Harris County Flood Control District create a map displaying a comparison of the ground elevations based on LiDAR and Base Flood Elevations (BFE) from the new 2004 Preliminary FEMA Floodplains. The map was created for educational purposes to help citizens determine if their property would benefit from an Elevation Certificate, which could result in savings of hundreds of dollars each year in flood insurance.

Cells shown in red on the map represent a ground elevation that is at least one foot above the BFE; a structure located in such an area would benefit from an Elevation Certificate. A light-green area on the map represents zero to six inches below the BFE. Depending on the type of construction and a structure's slab elevation, an Elevation Certificate would probably result in a reduction in the cost of flood insurance.

Courtesy of Harris County Flood Control District, Texas.

GreenInfo Network
San Francisco, California, USA
By Amanda Recinos

Contact
Larry Orman
larry@greeninfo.org

Software
ArcGIS Desktop 8

Printer
HP Designjet 1055cm

Data Source(s)
USGS, U.S. Census Bureau, GreenInfo Network, Los Angeles County

GreenInfo Network created a series of maps for The City Project to illustrate the location of parks and open space in relation to the neighborhood demographic attributes of poverty levels, percentage of youth, race/ethnicity, and access to a car. Each of these attributes was mapped individually to highlight any potential injustices based on a single factor. The 2000 Census tracts in the county were categorized to be above, at, or below the county average in each of the four variables. The tracts that were above the county average in all four categories were flagged as extremely disadvantaged (shown in red). Tracts with all four variables below the county average were marked as advantaged (shown in yellow). All remaining tracts were identified as being in range of the county average (shown in cream). The extremely disadvantaged and advantaged tracts were mapped alongside existing parks and open space.

In addition to the spatial analysis, park acres per thousand residents were calculated for cities and political districts. The acres-per-thousand-residents measure is a standard method of assessing a region's park infrastructure. Results revealed a wide range of park acreage per thousand residents with some of the healthiest cities having well over 400 acres per thousand residents, while some impoverished communities had less than one acre per thousand residents. As a whole, Los Angeles County is park-poor with 101 of the 131 analyzed communities falling below the national average of 6–10 acres of parks per thousand residents.

California Energy Commission Energy Facilities Siting and Environment

Public Interest Energy Research (PIER) Program
Sacramento, California, USA
By Lian Duan

Contact

Dora Yen
dyen@energy.state.ca.us

Software

ArcGIS Desktop 9.1

Data Source(s)

Project specific, CDF, NREL, CEC, HCD, USGS, DWR

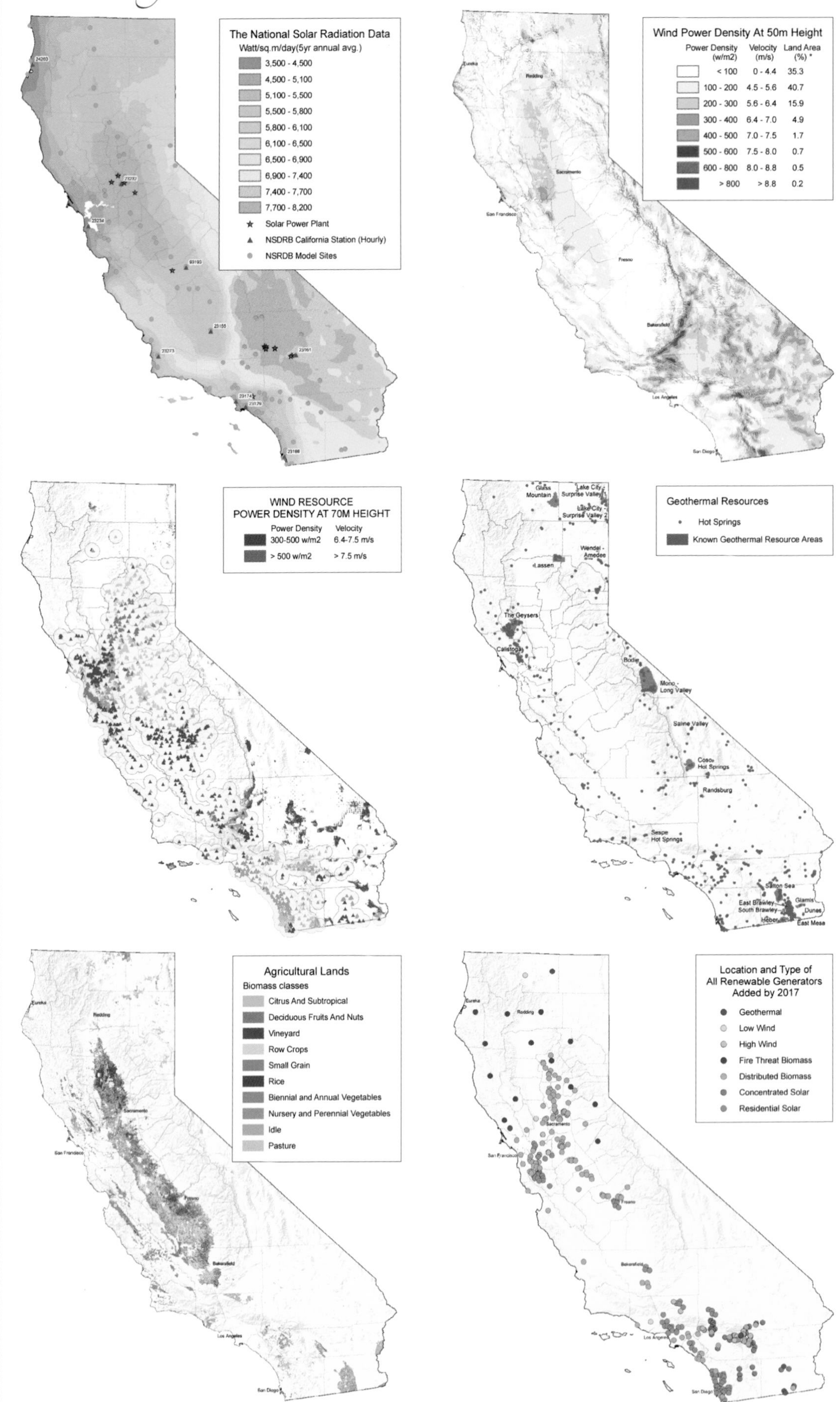

This collection of California maps represents a subset created to aid analysis as part of the Strategic Value Analysis (SVA) project. Developed by the California Energy Commission's Public Interest Energy Research (PIER) program, SVA was envisioned to strategically identify and locate new renewable generation areas to meet the state's Renewable Portfolio Standard goals. SVA presents a systems approach to maximizing the value of investment in renewable resources and, at the same time, addresses transmission system development needs statewide. The approach is noteworthy in that it quantifies the value of renewable energy resources and technologies at specific locations by factoring the value these facilities would bring to the transmission and distribution grid. Coupling advanced GIS analysis and visualization and a variety of data layers, including renewable resource assessments, land use, and transmission and load forecasts, a portfolio of renewable projects corresponding to locations with transmission benefit was developed.

Courtesy of the California Energy Commission.

City of Las Vegas
Las Vegas, Nevada, USA
By Jorge Morteo, Richard Wassmuth, and Mark House

Contact
Jorge Morteo
jmorteo@lasvegasnevada.gov

Software
ArcGIS Desktop, ArcEditor, ArcGIS 3D Analyst, SketchUp

Printer
HP Designjet 1055cm

Data Source(s)
City of Las Vegas, Clark County, U.S. Census

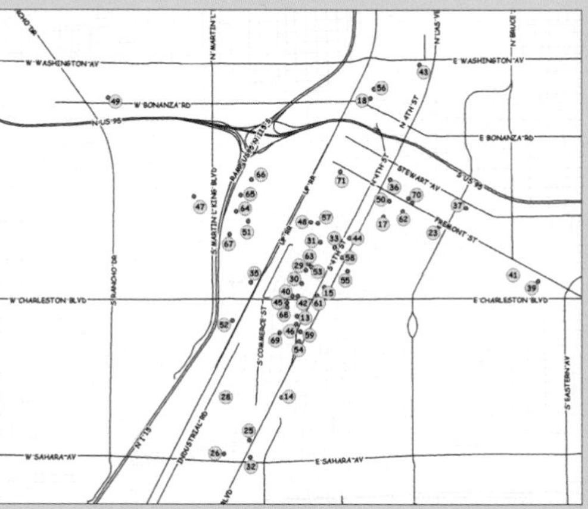

Location Map

Las Vegas added more than 27,000 new housing units between 1999 and 2004—basically, 2.13 new residents every hour, around the clock, and a new housing unit every hour and 36 minutes. The rapid population growth and the demand for living space have created an explosion of downtown mixed-use residential development. City authorities have been mapping this development within the downtown and surrounding areas and collecting attribute data such as the square footage of buildings, number of units, number of stories, height, and parking space for analysis purposes. In addition, projects that are proposed, approved, under construction, and completed are shown on the map. Several 3D GIS models have been assembled that show the height of buildings, streetscapes, historic places, and many other features.

All of this was accomplished using ArcGIS 3D Analyst and Site Builder (GIS Urban Simulator) ArcMap extensions. A powerful feature of the 3D models is that they are interactive, meaning users can travel inside them and see the impact that a proposed development or streetscape might have within the existing environment. This enhances public participation, and resolves problems before developments happen and thousands, or even millions, of dollars are spent on a project. In addition, the SketchUp plug-in for ArcGIS was used to created custom 3D features. This plug-in enables users to quickly create highly detailed three-dimensional models and seamlessly transfer them to an ArcGIS geodatabase for analysis. Finally, shadow studies are performed to see how shadows of buildings might impact an existing neighborhood at different times of the day and year.

Courtesy of the City of Las Vegas, Nevada.

July 1, 6:30 p.m.

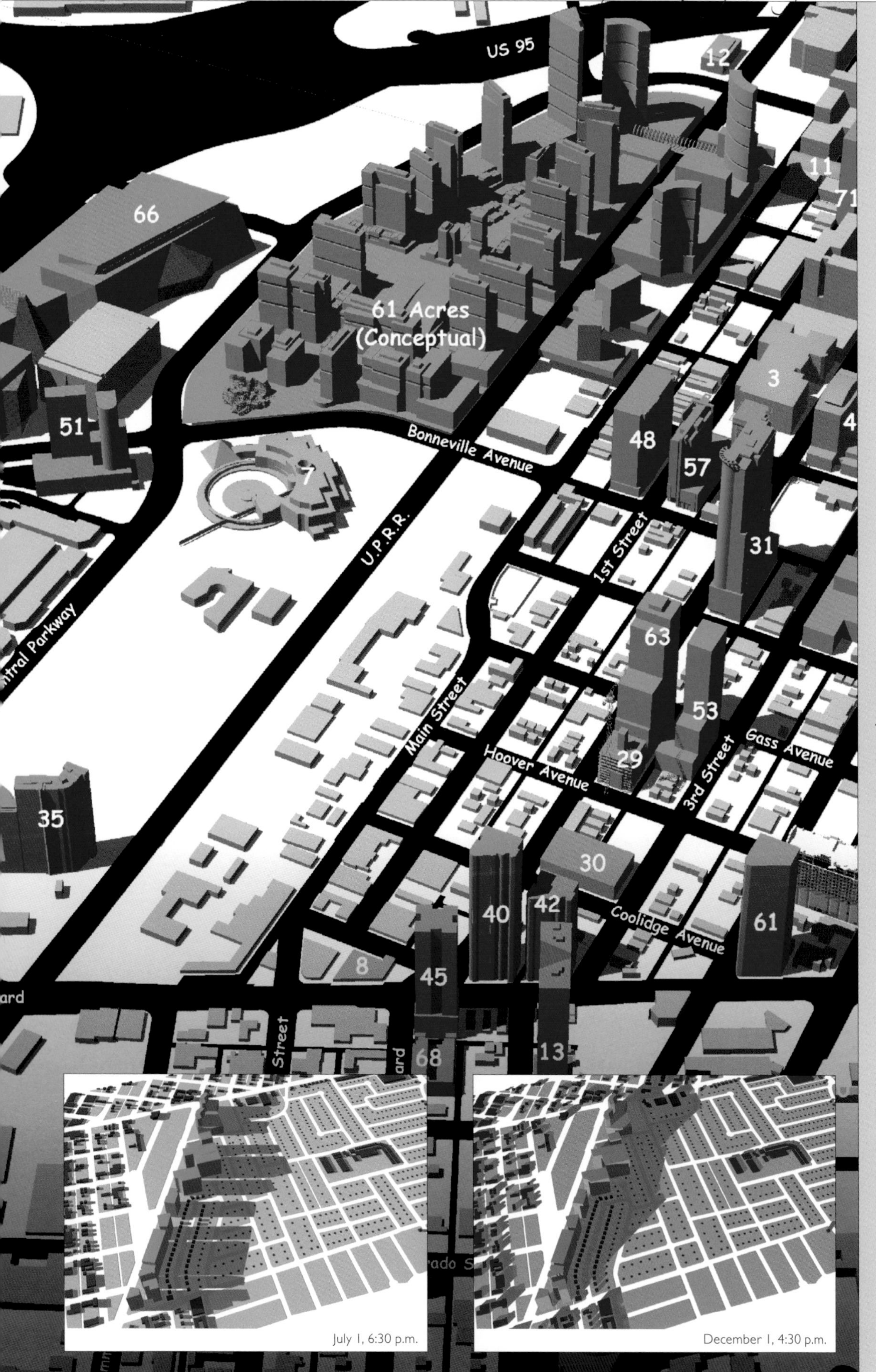

December 1, 4:30 p.m.

Carson Valley Employment and Housing

Parsons Corporation
San Jose, California, USA
By Eric Coumou

Contact
Eric Coumou
eric.coumou@parsons.com

Software
ArcGIS Desktop

Printer
HP Designjet 1055cm

Data Source(s)
State of Nevada, GDT

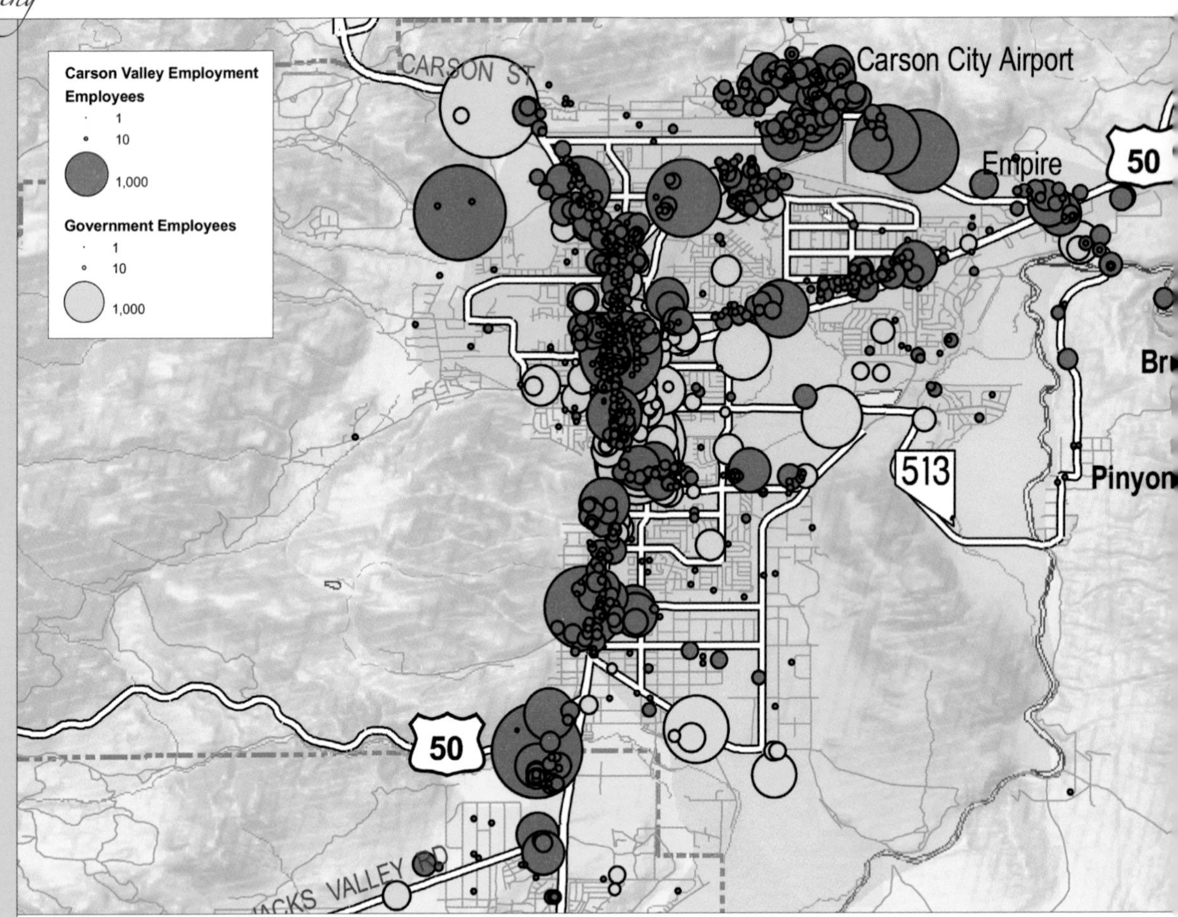

Carson Valley Employment

This map shows the distribution of employment in the Carson River Valley of Nevada. Each employer is represented with a dot whose size is proportional to the number of employees at that site. The number of employees was determined from federal employment records and augmented with surveys of the sites themselves. Government and school employment was further refined using state payroll information. The distribution of the employment centers and the number of employees, along with the distribution of residential dwelling units, is being used to update Traffic Analysis Zones (TAZs). Traffic Analysis Zones provide input to traffic models that will be used to identify problem areas and bottlenecks, and to design solutions.

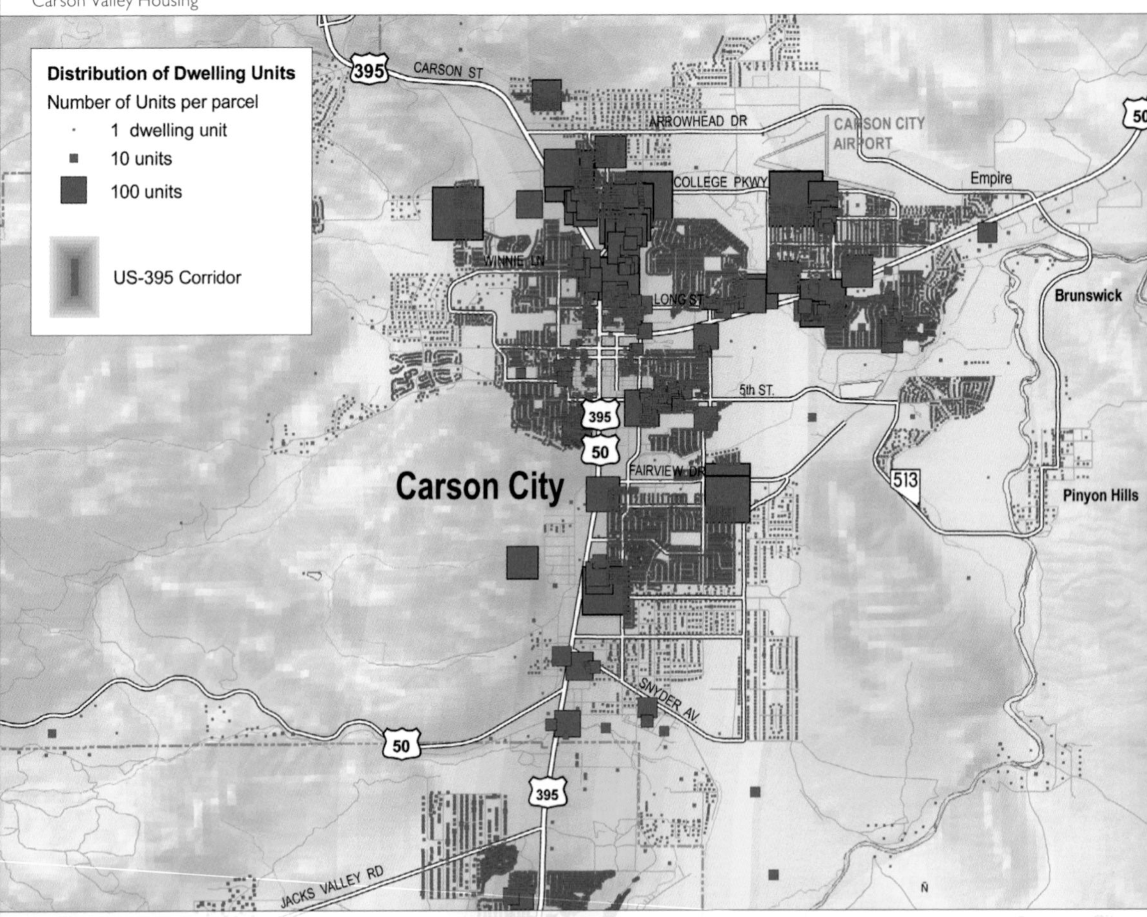

Carson Valley Housing

This map shows the distribution of housing in the Carson River Valley. Parcel data was acquired from the Douglas County and Carson City planning departments. Each parcel is represented with proportional-sized symbols showing the number of dwelling units on that parcel. Together with the Carson Valley Employment map, this map shows traffic planners specifically where the valley's residents are going each morning. Traffic models take these inputs and calculate traffic volumes and identify traffic bottlenecks. Proposed developments are added to the map to show proposed housing and projected traffic patterns. Traffic planners can then use these maps to identify potential future traffic problems and to design solutions.

Courtesy of Parsons Corporation.

WilsonMiller, Inc.
Fort Myers, Florida, USA
By Bryan J. Piersol, GISP; Chip McElroy; and
WilsonMiller, Inc.—Southwest Florida GIS Business Unit

Contact
Bryan J. Piersol, GISP
bryanpiersol@wilsonmiller.com

Software
ArcGIS Desktop 9.2, Adobe Photoshop

Plotter
Cannon iPF700

Data sources
Charlotte, Glades, Lee, Hendry, and Collier counties and their respective Property Appraisers Offices; Florida Department of Transportation; Florida Department of Environmental Protection; Southwest Florida Regional Planning Council; Bureau of Indian Affairs; USGS; ESRI; internal data developed by WilsonMiller, Inc.

The Geographical Regional Overview (GRO) map of Southwest Florida is a 1:135,000-scale representation of existing land use. It is a composite of a series of individual county GRO maps (Charlotte, Glades, Lee, Hendry, and Collier counties) created by the GIS Business Unit of WilsonMiller, Inc. The parcels are shaded based on the existing Florida Department of Revenue tax assessment codes assigned to each parcel. These individual maps are updated annually to satisfy the needs of land investors and developers. The map shows where development is currently taking place, and identifies vacant lands that are available for future development.

Courtesy of WilsonMiller, Inc.

Exploring Landfill Gas Production with GIS

Sanborn, Head & Associates, Inc.
Akron, Ohio, USA
By Daniel Schweitzer

Contact

Daniel Schweitzer
dschweitzer@sanbornhead.com

Software
ArcGIS Desktop, ArcGIS Spatial Analyst, ArcGIS 3D Analyst, Adobe Illustrator

Printer
HP Designjet D800

Data Source(s)
Topographic maps, CAD linework, field-collected data

Landfill A: Extraction Well Efficiency, November 2005
2D Bubble Plot

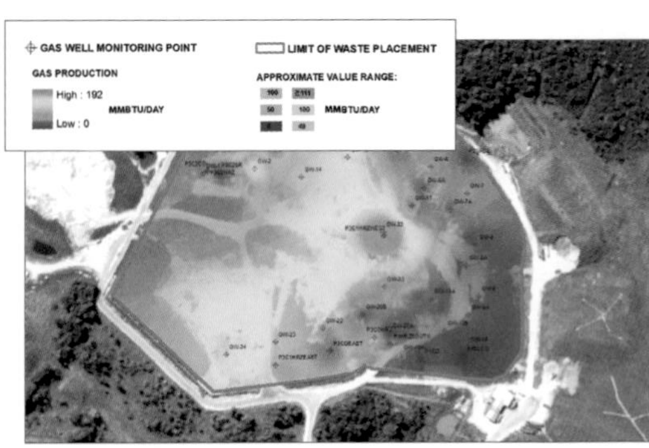

Landfill A: Extraction Well Efficiency, November 2005
Color Ramp Normalized to 111MMBTU/Day

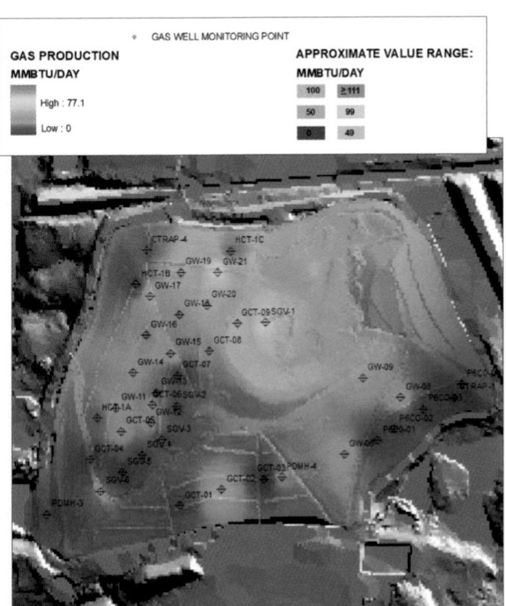

Landfill B: Gas Production (MMBTU/Day), February 15–28, 2006

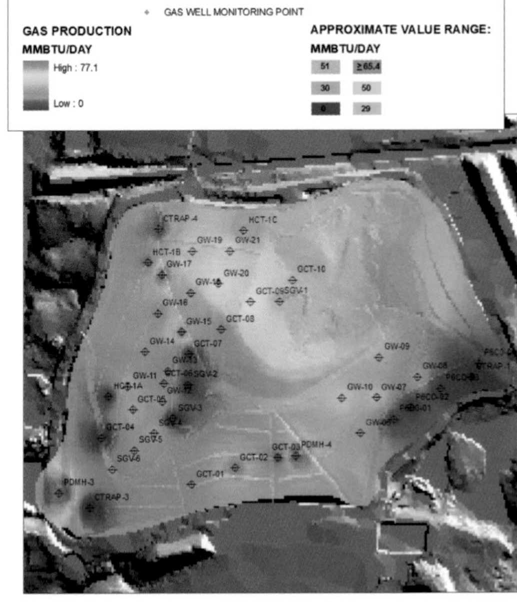

Landfill B: Gas Production (MMBTU/Day), December 1–15, 2005
Standardized to Average Maximum Monthly Production

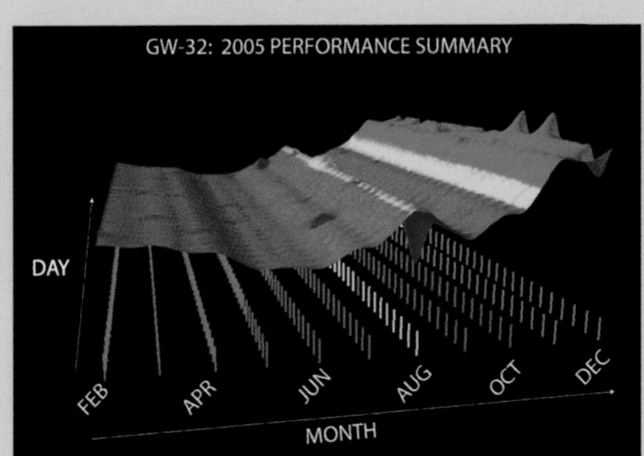

Who says GIS can't do time-series graphs?

Sanborn, Head & Associates, Inc. (SHA), of Akron, Ohio, uses GIS to incorporate regularly collected landfill tuning data into progress prints used by its engineers and clients to evaluate landfill conditions. Landfill tuning data includes measurements of methane production, gas temperature, and gas well flow rates, which are important indicators of a landfill's "health." Individual tuning rounds are plotted in separate data frames in the same document, which allows an engineer or client to quickly discern whether conditions are changing (or have changed). The visually intuitive framework that SHA is able to create with GIS has been very well received by its clients.

Courtesy of Daniel Schweitzer, Sanborn, Head & Associates, Inc.

Monthly Snapshots at Landfill A

February 2005

April 2005

June 2005

August 2005

October 2005

December 2005

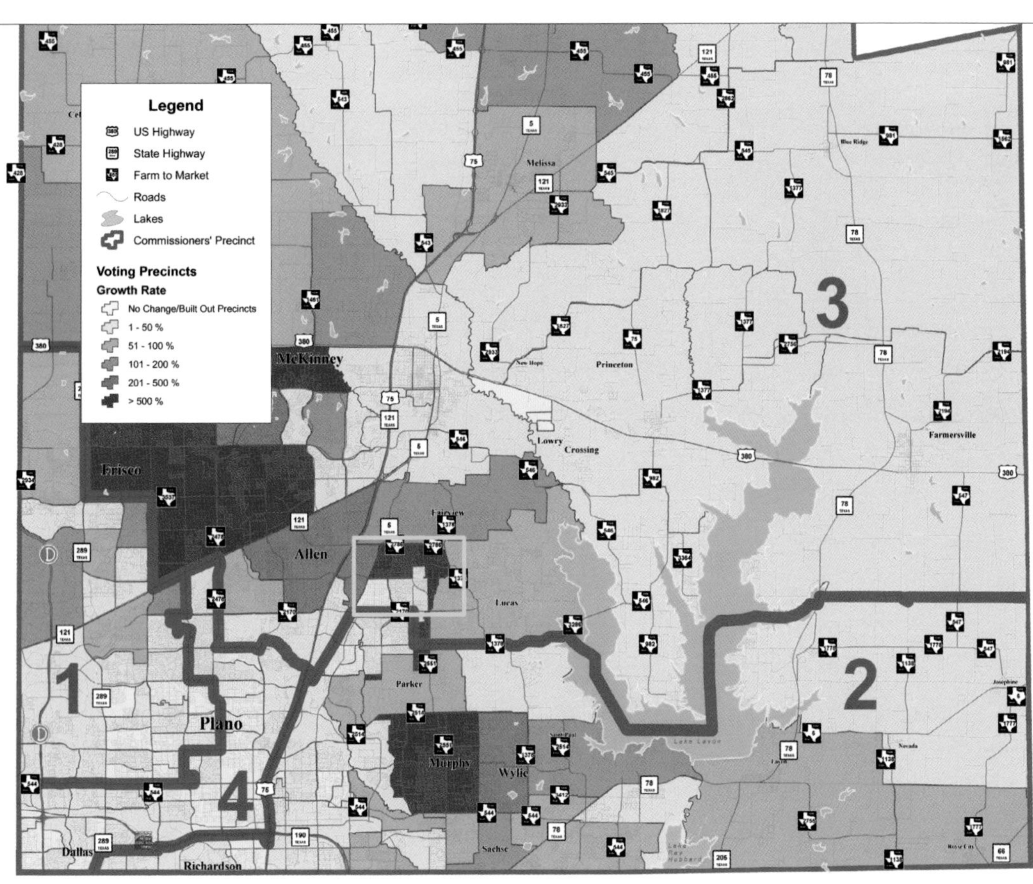

Estimated Population Growth Rate by Precinct (2000–2006)

Parcel Growth Density (2000–2006)

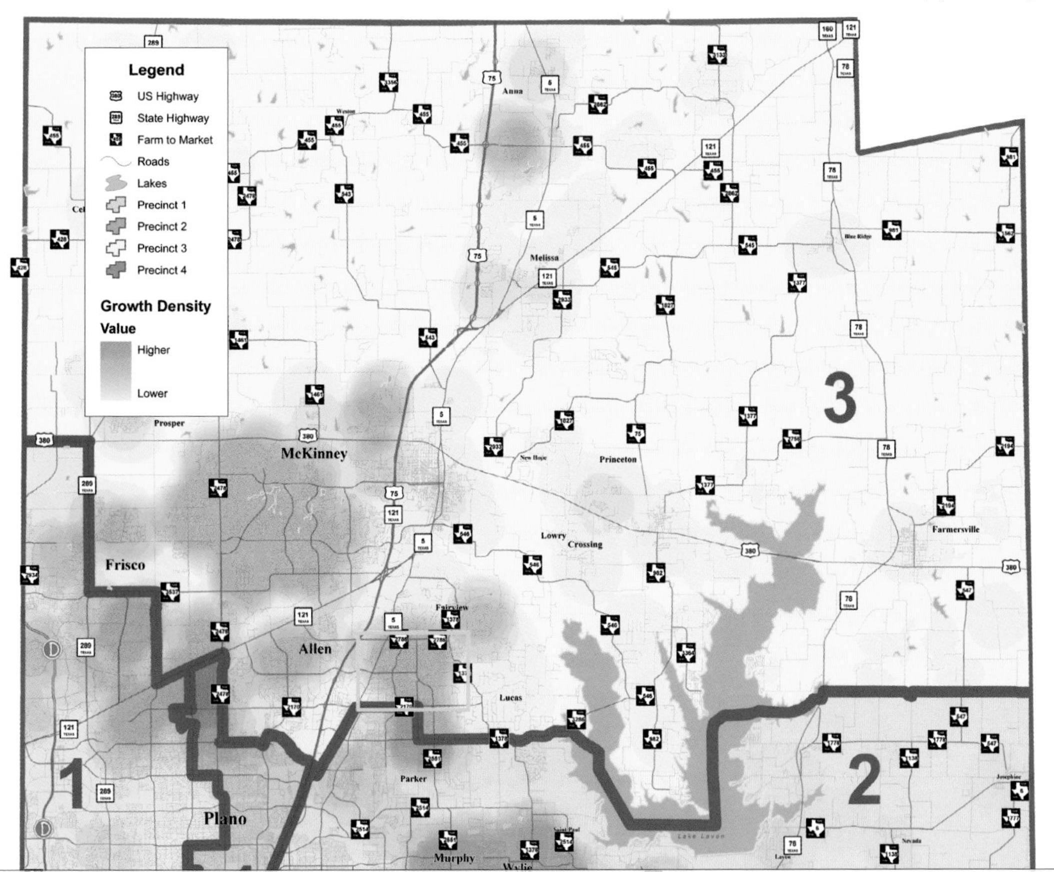

Collin County

McKinney, Texas, USA

By Nick Enwright, Tim Nolan, Kendall Holland, Ramona Luster, Bret Fenster, John Kinnaird, and Russ Frith

Contact

Bret Fenster

bfenster@collincountytx.gov

Software

ArcGIS Desktop 9.1

Printer

HP Designjet 5500uvps

Data Source(s)

U.S. Census 2000, NCTCOG 2006 population estimates, Collin County GIS

2001 Voting Precinct #6

2005 Voting Precinct #6

Fueled by six of North Central Texas' fastest-growing cities, Collin County is one of the state's fastest-growing counties. The objective of this study was to use GIS to demonstrate a methodology for assessing the need for county redistricting. The population for Collin County was obtained from the U.S. Census for 2000, and the January 1, 2006, projected county population was provided by the North Central Texas Council of Governments. New residential construction data was obtained from the Central Appraisal District of Collin County. The projected population was assigned to each precinct in direct proportion to the increase in new construction.

Evident on this map are growing pains accompanying the county's intense growth in its attempt to maintain balanced populations for each commissioner's district. Population growth from 2000 to 2006 has not increased uniformly throughout the county, resulting in severely imbalanced populations among districts.

Courtesy of Collin County GIS.

New Developments Map

City of Indio

Indio, California, USA
By Emina Sulych

Contact

Emina Sulych
esulych@indio.org

Software

ArcGIS Desktop 9, AutoCAD 2006, Adobe Photoshop

Printer

HP Designjet 5000

Data Source(s)

City of Indio, Riverside County

STREET IMPROVEMENTS
CAPITAL PROJECTS
JOIN AGENCY PROJECTS
DEVELOPERS IMPROVEMENTS
Legend

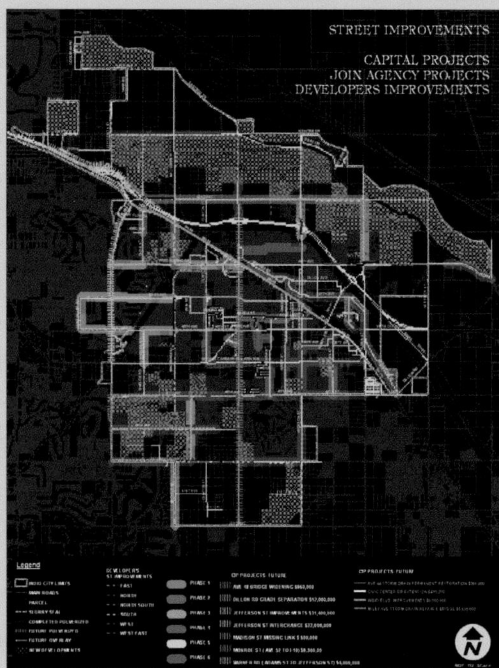

The New Developments Map provides critical information on new developments within the city of Indio. Indio's population has increased 42 percent over the last five years, with an average annual growth rate of more than 8 percent. This translates to more than 6,500 new rooftops with an additional 3,000 forecast for the next three years. The map provides up-to-date information to municipal departments on the location of these new developments, as well as a basis for future infrastructure planning.

The creation of the New Developments Map was spawned by enormous growth in residential and commercial areas of the city. This map served as a basemap for future planning of residential and commercial sites, street, water, sewer improvements, and other infrastructure planning needs. Information provided has been utilized by several city departments, developers, agencies, and residents and proven crucial for a variety of forecasts and projections.

Courtesy of the City of Indio, California.

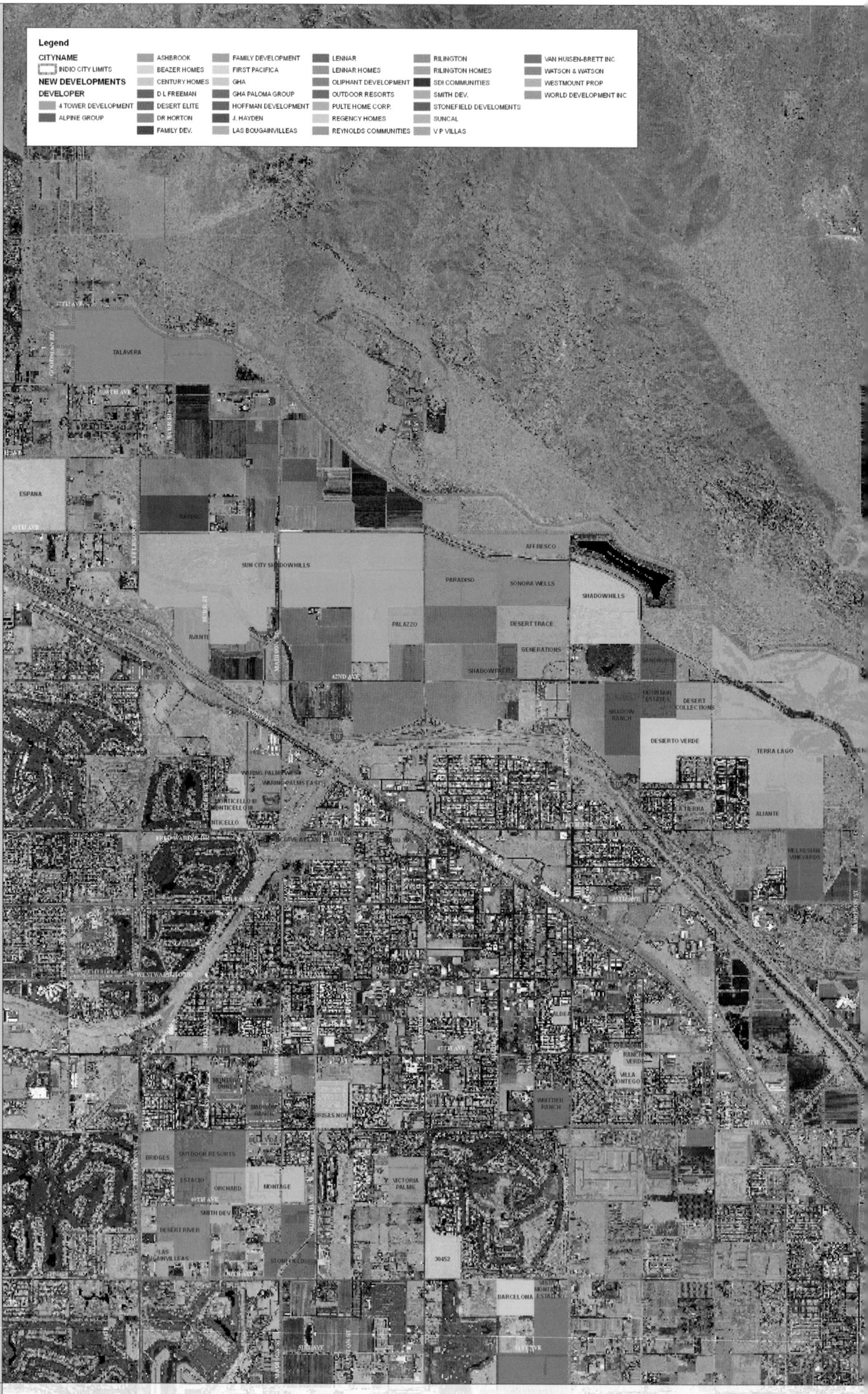

Legend
CITYNAME
INDIO CITY LIMITS
NEW DEVELOPMENTS
DEVELOPER
4 TOWER DEVELOPMENT
ALPINE GROUP
ASHBROOK
BEAZER HOMES
CENTURY HOMES
D L FREEMAN
DESERT ELITE
DR HORTON
FAMILY DEV.
FAMILY DEVELOPMENT
FIRST PACIFICA
GHA
GHA PALOMA GROUP
HOFFMAN DEVELOPMENT
J. HAYDEN
LAS BOUGAINVILLEAS
LENNAR
LENNAR HOMES
OLIPHANT DEVELOPMENT
OUTDOOR RESORTS
PULTE HOME CORP.
REGENCY HOMES
REYNOLDS COMMUNITIES
RILINGTON
RILINGTON HOMES
SDI COMMUNITIES
SMITH DEV.
STONEFIELD DEVELOMENTS
SUNCAL
V P VILLAS
VAN HUISEN-BRETT INC
WATSON & WATSON
WESTMOUNT PROP
WORLD DEVELOPMENT INC
TALAVERA
ESPANA
SUN CITY SHADOWHILLS
AFFRESCO
PARADISO
SONORA WELLS
SHADOWHILLS
PALAZZO
DESERT TRACE
AVANTE
GENERATIONS
42ND AVE
DESERT COLLECTIONS
DESIERTO VERDE
TERRA LAGO
ALIANTE
BARCELONA
VICTORIA PALMS
MONTAGE
ORCHARD
BRIDGES
DESERT RIVER
30452

How Far Is It to Get Milk? Access to Major Grocery Stores for the Poverty-Level Population in Rhode Island

Environmental Data Center
Kingston, Rhode Island, USA
By Mark Christiano, Dr. Nancy Fey Yensan, Monica Belyea, and Lorraine Keeney

Contact
Mark Christiano
mark@edc.uri.edu

Software
ArcGIS Desktop 9.1

Printer
HP Designjet 500ps

Data Source(s)
RIGIS, U.S. Census, Providence Plan

Massachusetts

Grocery Stores
Community Health Centers
Bus Routes
30 Minute Travel Time

WOONSOCKET
CUMBERLAND
NORTH SMITHFIELD
LINCOLN
SMITHFIELD
CENTRAL FALLS
PAWTUCKET
NORTH PROVIDENCE
PROVIDENCE
SCITUATE
JOHNSTON
EAST PROVIDENCE
CRANSTON
BARRINGTON
WARREN
WARWICK
BRISTOL
WEST WARWICK
EAST GREENWICH
TIVERTON
PORTSMOUTH
EXETER
JAMESTOWN
NORTH KINGSTOWN
MIDDLETOWN
NEWPORT
SOUTH KINGSTOWN
NARRAGANSETT

For many people living at or below the poverty level, transportation to a grocery store is a challenge that limits a household's ability to get food. Many of the larger grocery stores that can offer lower prices and better selection are located in more affluent suburban neighborhoods. Urban and rural low-income households that cannot afford a vehicle must use public transportation, walk, or pay for a ride to get to larger stores. Using ArcGIS Network Analyst, the Environmental Data Center, a GIS laboratory in the University of Rhode Island's Department of Natural Resources Science, modeled areas that are within 30 minutes of a major grocery store if a person had to walk or use public transportation. The resulting data was then compared with U.S. Census poverty data to determine regions in Rhode Island where people have trouble accessing grocery stores that offer greater variety and lower prices.

Courtesy of Mark Christiano, Environmental Data Center.

A GIS Decision Model for Detecting Substandard Housing

Alachua County, Florida

Gainesville, Florida, USA

By GIS Division, Department of Growth Management, Alachua County Board of Commissioners

Contact

Juna Goda Papajorgji

jpapajorgji@alachuacounty.us

Software

ArcGIS Desktop 9.1

Printer

HP Designjet 800

Data Source(s)

U.S. Census 2000, Alachua County, FGDL

This is a decision support model developed by the University of Florida in support of activities of the Alachua County Housing Authority and the Alachua County Board of Commissioners. The model identifies deteriorated housing stock, classifying it by intensity of deterioration into the following categories: High Substandard, Medium Substandard, Low Substandard, and Suspected Substandard. The model also determines a deterioration gradient per one-mile section grid of the Public Lands Survey System, based on a number of substandard and suspected structures per section. A customized interface provides for mapping at a parcel level and for generating address lists by section grid. The model can be used for identification of areas of increasing property structure deterioration and for detection of annual changes and trends in substandard housing. The model is not proprietary.

Courtesy of Alachua County Board of Commissioners, Florida.

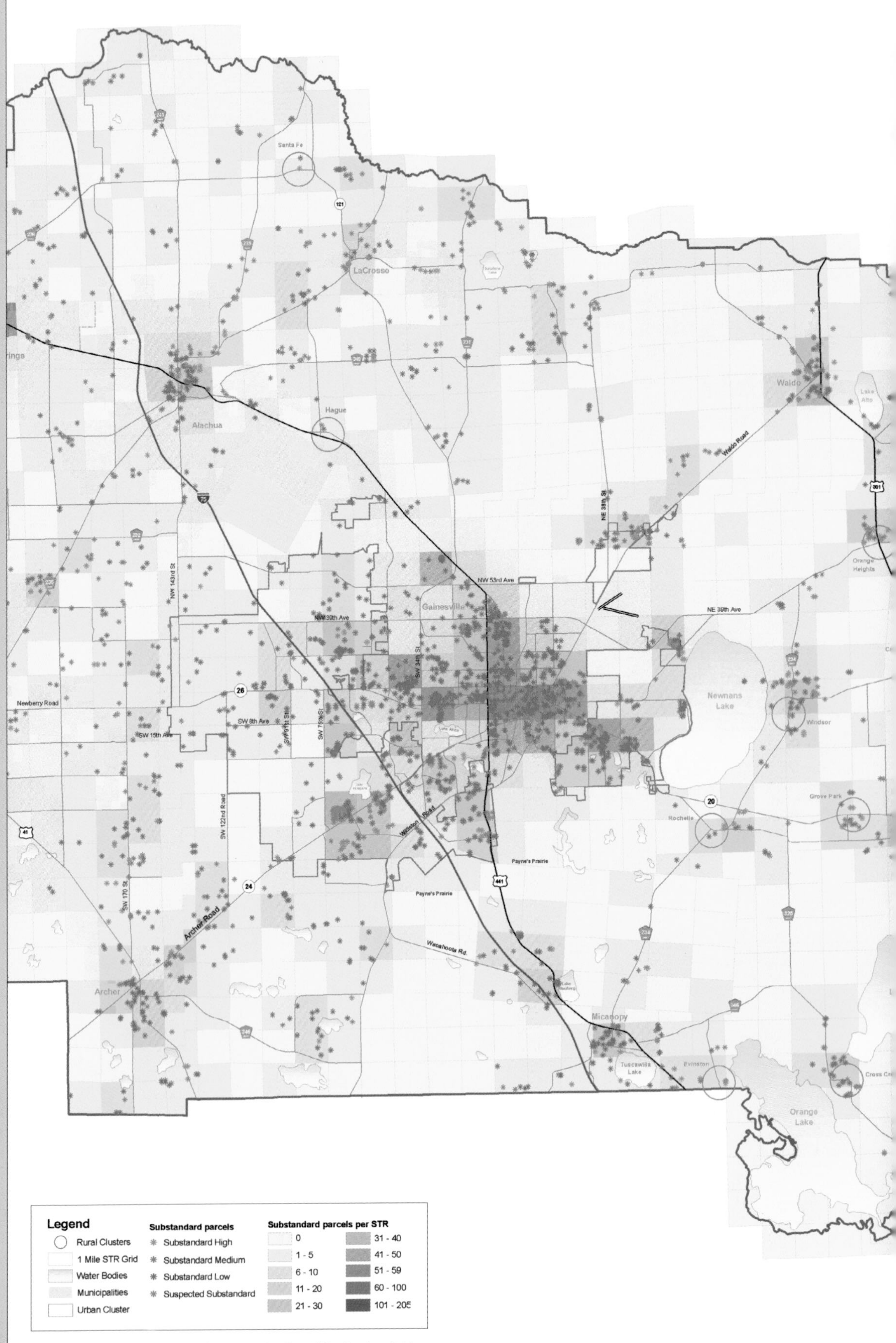

Concentration of Substandard Structures by One-Mile Section Grid

Health and Human Services Master Plan 2005–2015

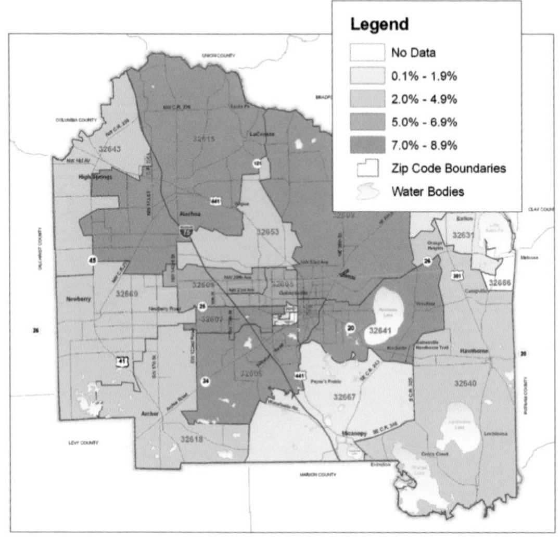

Percentage of Residents Receiving Services at the Department of Community Support Services, Categorized by ZIP Code

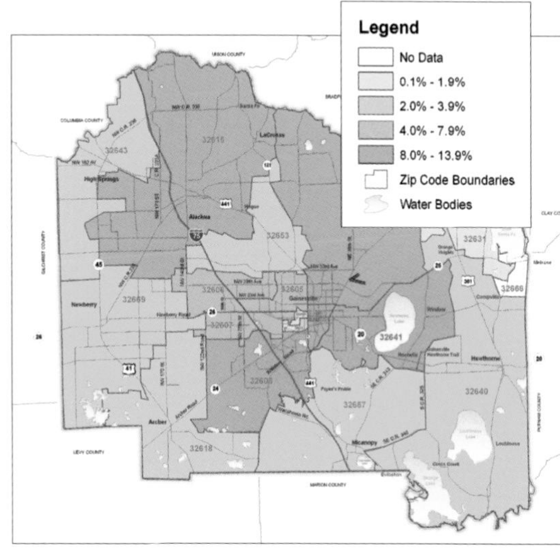

Percentage of Persons Calling the Department of Community Support Services for Information, Categorized by ZIP Code

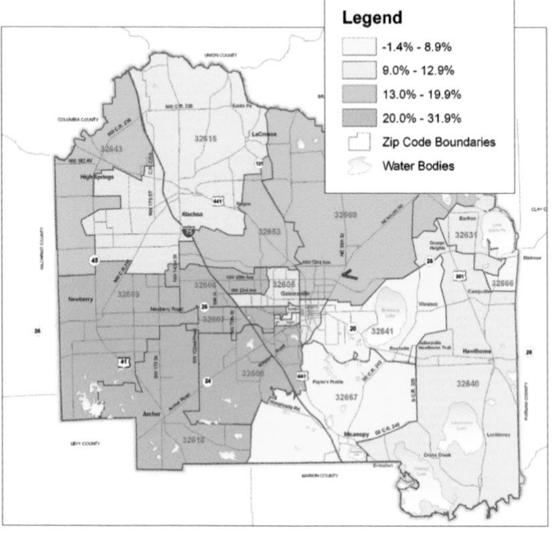

Percentage Change in Population 2000–2009, Categorized by ZIP Code

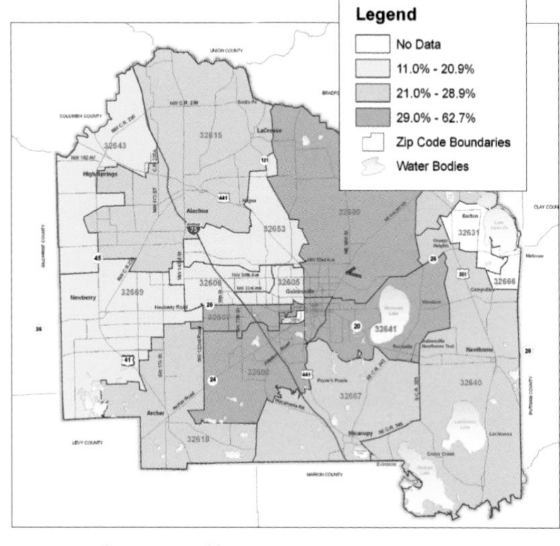

Percentage of Persons Living at or Below 150% of the 1999 Federal Poverty Level, Categorized by ZIP Code

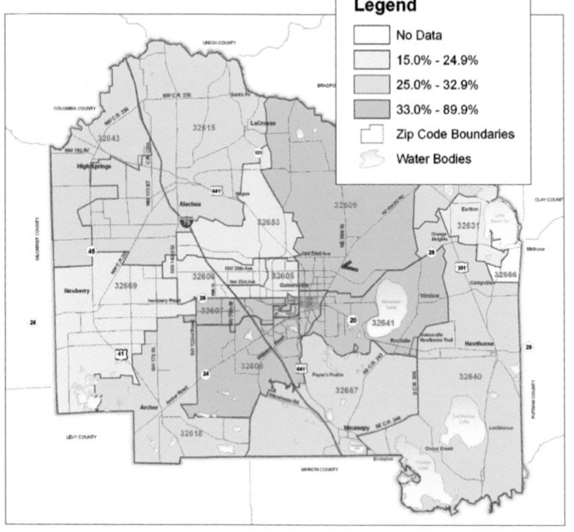

Percentage of Households with an Income Level Less than $25,000, Categorized by ZIP Code

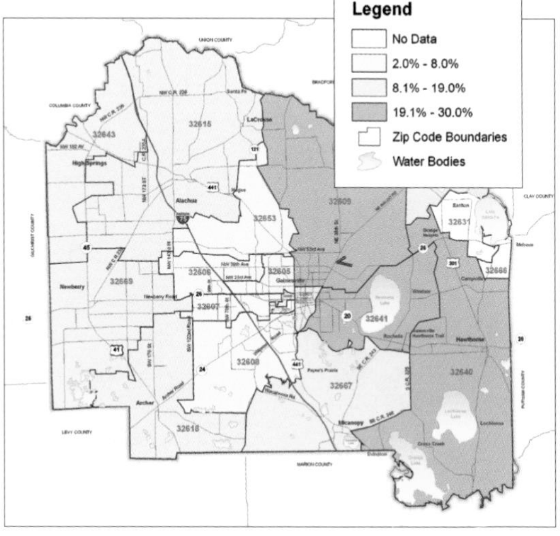

Percentage of Persons 25 years or older without a High School Diploma, Categorized by ZIP Code

Alachua County, Florida

Gainesville, Florida, USA

By GIS Division, Department of Growth Management, Alachua County Board of Commissioners

Contact

Juna Goda Papajorgji

jpapajorgji@alachuacounty.us

Software

ArcGIS Desktop 9.1

Printer

HP Designjet 800

Data Source(s)

U.S. Census 2000, Alachua County, FGDL

The Health and Human Services Master Plan 2005 was prepared to assist the Department of Community Support Services (DCSS) in Alachua County, Florida. The plan identifies social service needs for the next decade, emphasizing coordination of health services with human services. To support the objectives of this plan, a GIS analysis was conducted. Social measurements derived from U.S. Bureau of the Census data were compared with DCSS service data for the last decade. The distribution of the social measurements revealed locations of the most needy residents and potential future allocation of population growth. The DCSS service data indicated that past and current service areas were evenly distributed throughout the county.

Courtesy of Alachua County Board of Commissioners, Florida.

Optimum Exposure Area Grid Placement for Human Health Risk Assessment Using ArcGIS Spatial Analyst

BWXT Pantex, LLC
Amarillo, Texas, USA
By Jeff Stovall and Michelle Bolwahnn (BWXT Pantex)
and David Profusek (SAIC)

Contact
Jeff Stovall
jstovall@pantex.com

Software
ArcGIS Desktop 8.3, Adobe Photoshop CS2

Printer
HP DesignJet 800

Data Source(s)
Site data

The Environmental Restoration Program for the U.S. Department of Energy/National Nuclear Security Administration's Pantex Plant recently completed a baseline human health risk assessment of exposure to hazardous substance releases from legacy waste management units to help determine the need for further cleanup actions.

A quantitative analysis method using the PointStats Neighborhood Analysis function of the ArcGIS Spatial Analyst extension was used to determine the optimum exposure grid placement for risk evaluation at the Pantex Plant. The objectives included evaluating the impact of various grid configurations on risk results and determining if risk results are sensitive to exposure grid configuration. The optimum grid placement was selected using the most sensitive grid configuration and by locating the grid in areas of elevated risk with consideration of worker patterns and facility use.

PointStats Neighborhood Analysis

The ArcGIS PointStats analysis tool is used to identify the shape and orientation of the neighborhood around a hot spot resulting in the highest average risk ratio when the grid is centered on that location. Six different configurations were assessed: a horizontal and vertical rectangle, a square, and each of those shapes turned at a 45-degree angle.

Exposure Area Grid Creation

A regular grid is then replicated around the beginning cell, using the same configuration as the beginning cell, so that the grid encompasses the entire area being assessed.

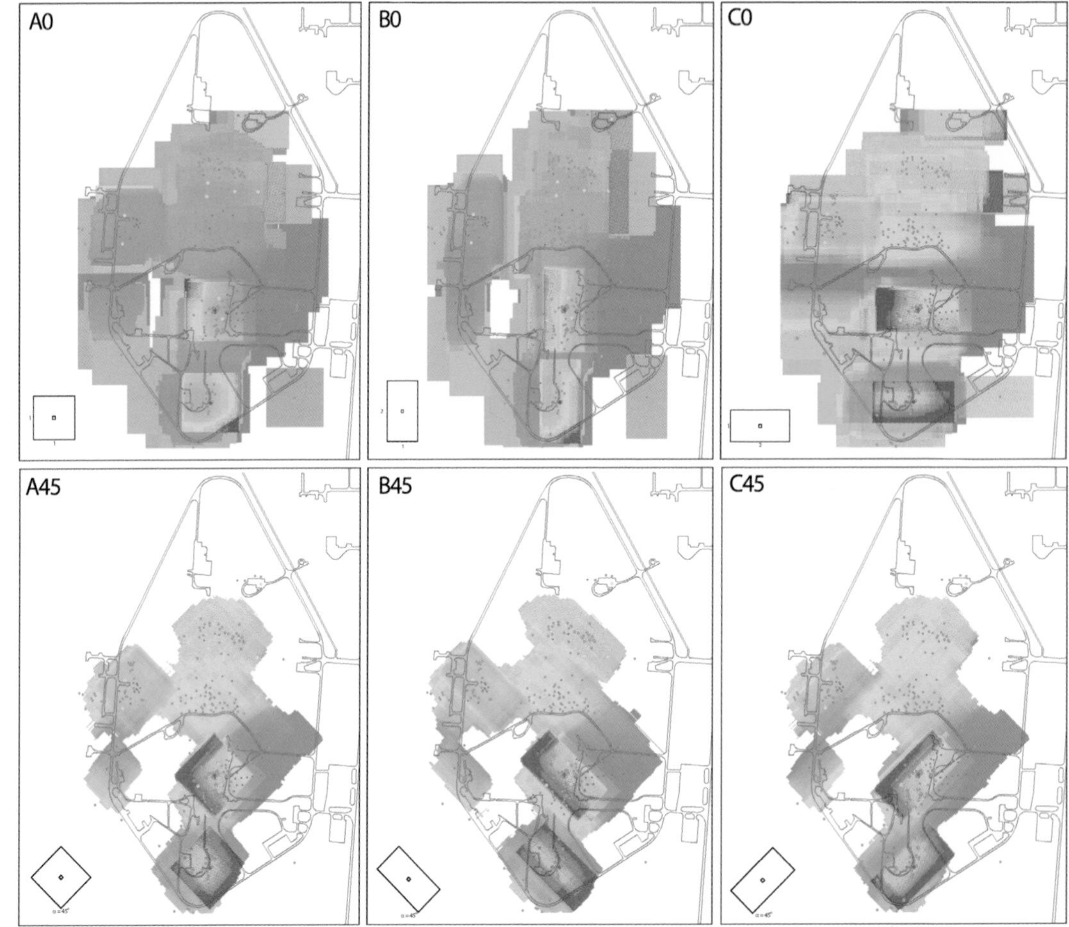

PointStats Neighborhood Analysis

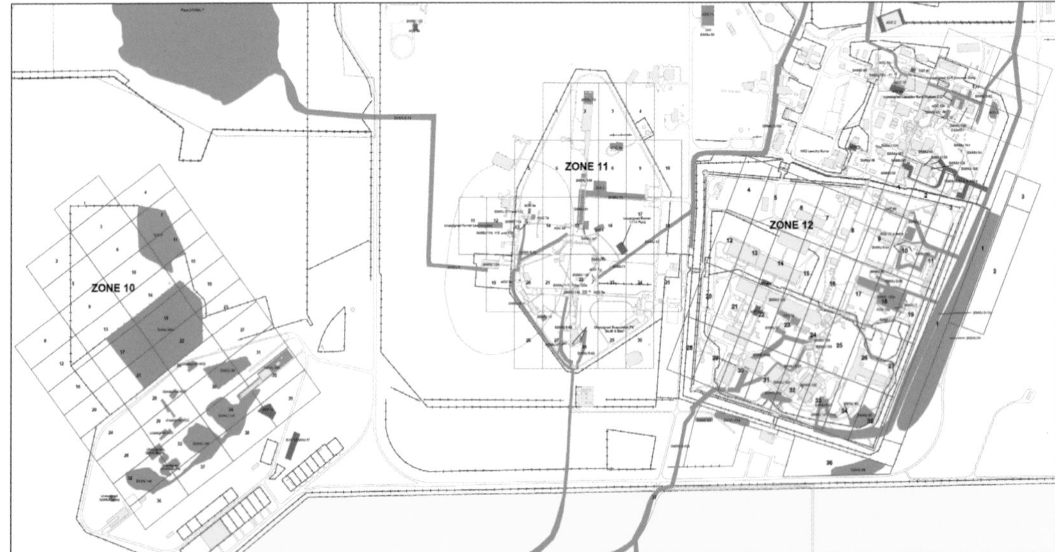

Final Exposure Area Grids

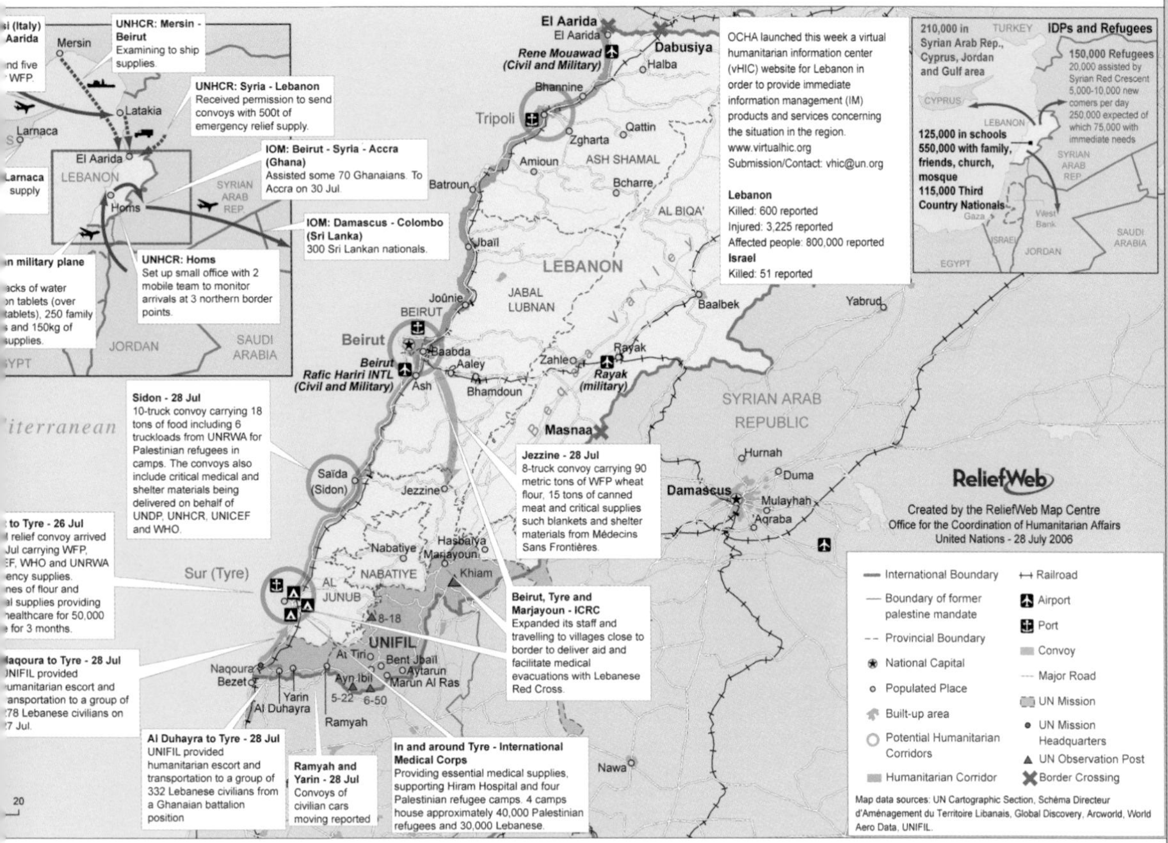

United Nations Office for the Coordination of Humanitarian Affairs

New York, New York, USA

By United Nations OCHA ReliefWeb Map Centre

Contact

Lorant Czaran

czaran@un.org

Akiko Harayama

harayama@un.org

Software

ArcGIS Desktop 9, Adobe Illustrator

Printer

HP Designjet 5550uvps

Data Source(s)

UN Cartographic Section, UN OCHA, ArcWorld

Lebanon Humanitarian Situation Map

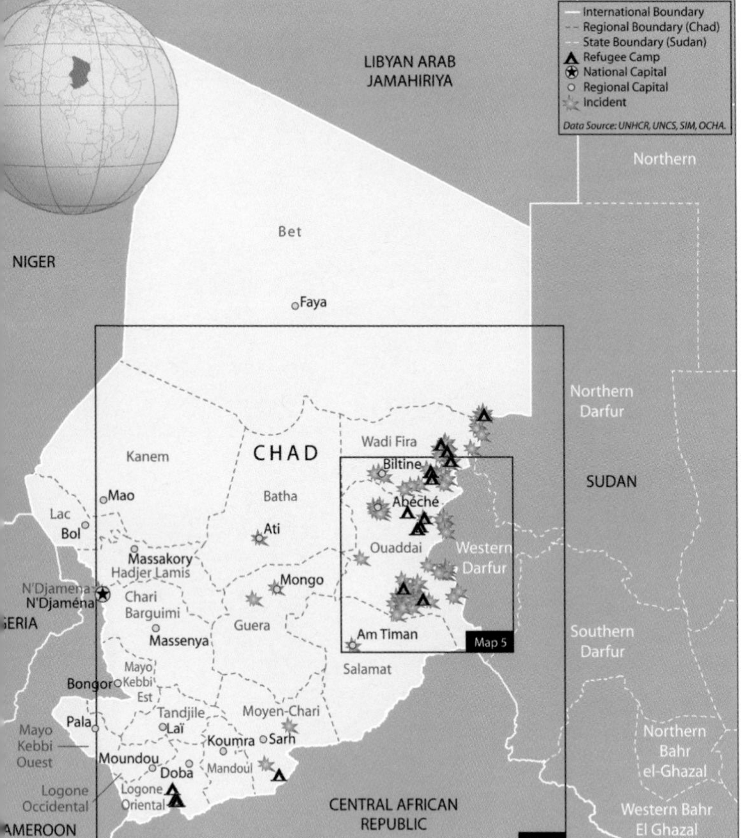

Chad Humanitarian Profile Map

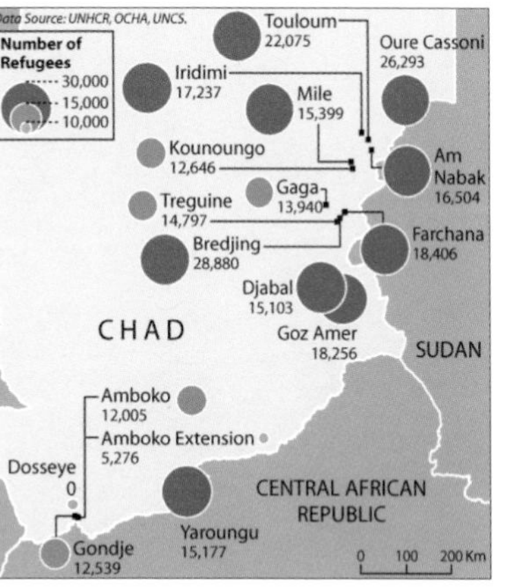

Map 4: Refugees (as of October 2006)

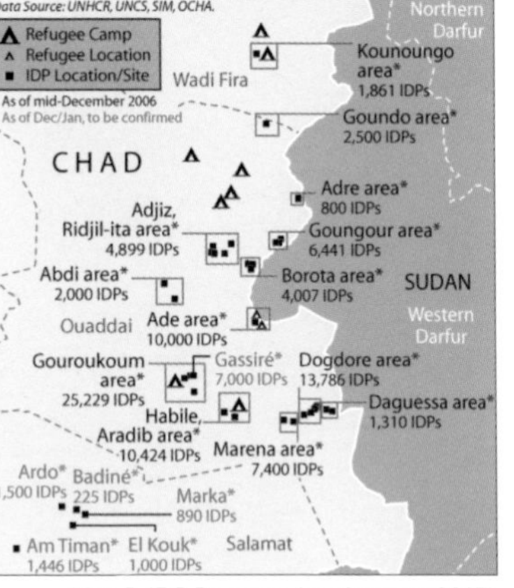

Map 5: IDP Presence (as of January 2007)

The Lebanon Humanitarian Situation Map was produced in August 2006 during the weeks of the cross-border crisis, as part of a situation map series linked to the Humanitarian Situation Reports released by the Office for the Coordination of Humanitarian Affairs (OCHA). Such public maps are core products of the ReliefWeb Map Centre in supporting the United Nations (UN) humanitarian response in case of conflicts or natural disasters; these map products are widely disseminated and often the first and very timely source for response and decision making during the early stages of a humanitarian emergency.

The Chad Humanitarian Profile Map is part of a relatively new set of products developed by the ReliefWeb Map Centre, and is aimed primarily for outreach as well as concise illustration of complex emergencies or natural disasters in countries or regions around the world. The production involves extensive data research and liaison with field offices and other UN agencies for access to the latest relevant information to be mapped or illustrated with well-targeted information graphics. It is a map product very well received by the UN and external partner users in the humanitarian community. Please visit www.reliefweb.int for more information.

Courtesy of United Nations Office for the Coordination of Humanitarian Affairs.

USAID/Nepal
Kathmandu, Nepal
By Indra S. KC

Contact
Amy Paro
aparo@usaid.gov

Software
ArcGIS Desktop 9

Data Source(s)
Nepal Survey Department, Government of Nepal, USAID mission in Nepal, and partners

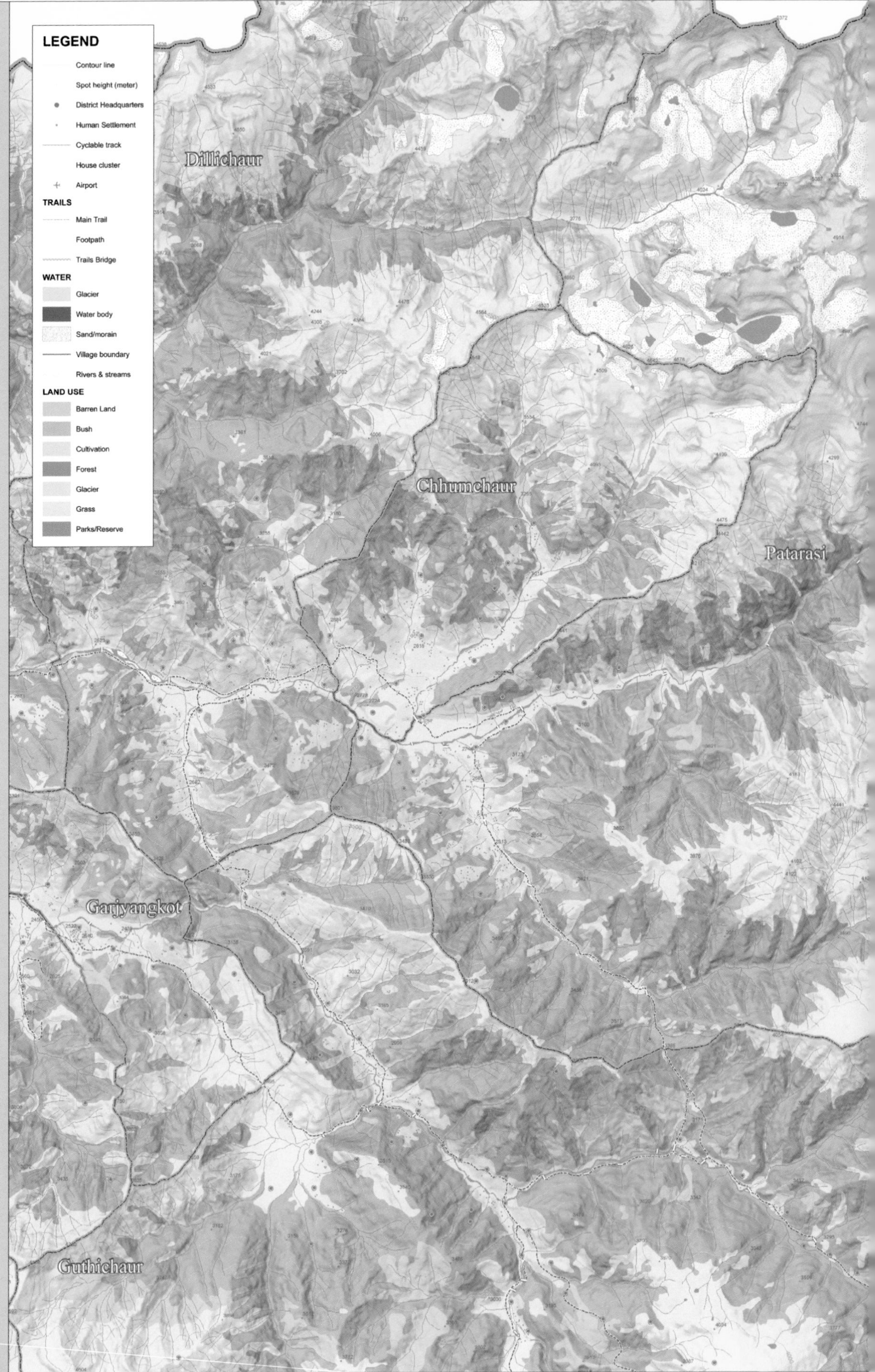

USAID/Nepal produced these two maps to aid the implementation of social programs created to improve living conditions and disaster preparedness in the remote Jumla district of Nepal by developing trails connecting villages and district headquarters. These maps were produced using topographic basemap data from the Government of Nepal's Survey Department and with available information from the mission and its partners. Analytical cartography tools and techniques were applied using ArcGIS to identify rural human settlements, trails, terrain, and land use.

Human Settlements and Land Use

The settlements in the Jumla district are scattered mainly on the slopes of mountains along river basins such as the Tila River. Most of these areas are at elevations below 2,700 meters, where cultivated lands make up about 12 percent of the district's total land use. A significant portion of land cover is made up of grasslands and highland meadows often used for grazing Himalayan goat, sheep, and yak. These lands are abundant in medicinal herbs and have great potential for sustainable economic development.

Since Jumla is not yet connected by roads, villagers and animals are used to move food, supplies, and sick people between settlements along trails and footpaths. In emergency situations like drought, people must walk days to collect provisions. The trail along the Tila River (shown in orange/red), leading from one end of the district to the district headquarters in Khalanga, is only wide enough for several bicycles. A tractor, assembled after its parts were brought in by airplane, is used to transport goods to and from the airport and the district headquarters.

Courtesy of USAID/Nepal.

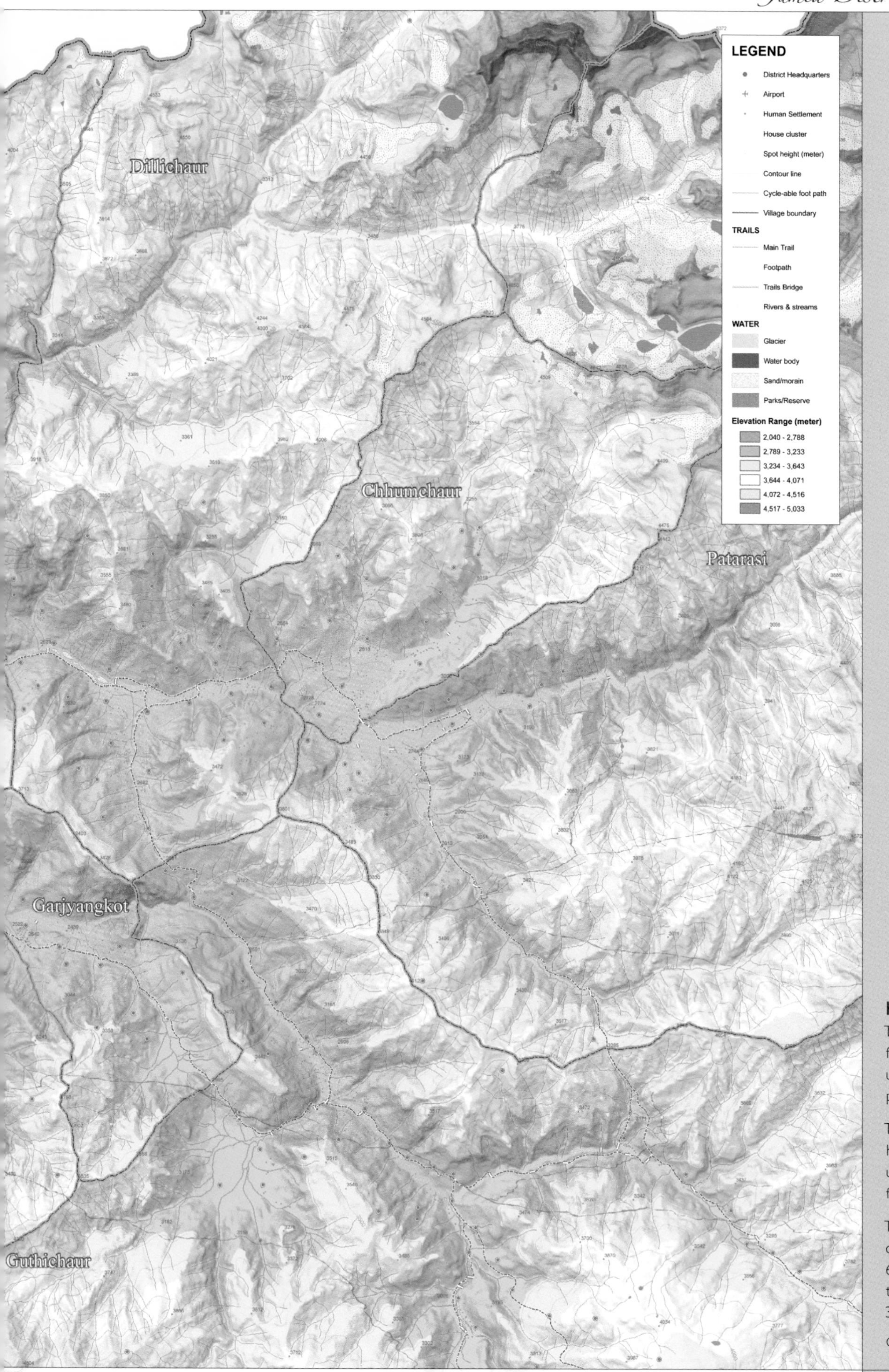

USAID/Nepal
Kathmandu, Nepal
By Indra S. KC

Contact
Amy Paro
aparo@usaid.gov

Software
ArcGIS Desktop 9

Data Source(s)
Nepal Survey Department, Government of Nepal, USAID mission in Nepal, and partners

Human Settlements and Terrain

These maps helped USAID/Nepal to identify trails and footpaths connecting villages and rural markets in order to upgrade them to animal tracks, which helps alleviate the perpetual suffering of people carrying loads on their backs.

This color-coded elevation map shows the distribution of human settlements and the network of trails that people use for transport and communication on a rugged land surface in the Himalayas.

The elevation in the Jumla district ranges from a minimum of 2,040 meters to a maximum of 6,600 meters. About 60 percent of land is within the elevation range of 2,040 to 3,600 meters. Only 40 percent of the land is above 3,600 meters.

Courtesy of USAID/Nepal.

Portland State University

Portland, Oregon, USA
By Richard Lycan

Contact

Richard Lycan
lycand@pdx.edu

Software

ArcGIS Desktop 9.1, ArcGIS Spatial Anayst, Macromedia FreeHand

Printer

HP Designjet 500ps

Data Source(s)

Vital statistics, U.S. Census, Portland Metro RLIS, enrollment data from Portland Public Schools

1. Births in 1995
2. Five Year Net Migration
3. KG Capture Rate
4. Five Year Retention
5. KG in 2000

1. Children are born.
2. Some move away and some move in. Red shows net gains, blue losses.
3. The numbers are reduced by the "capture rate," at the proportion attending public schools.
4. Migration multiplied times the capture rate equals five-year retention.
5. Yielding the number of students in kindergarten from births five years earlier.

The Progression from Birth to Kindergarten

The Analysis of Births by Age, Race, and Location Yields a Forecast of Births

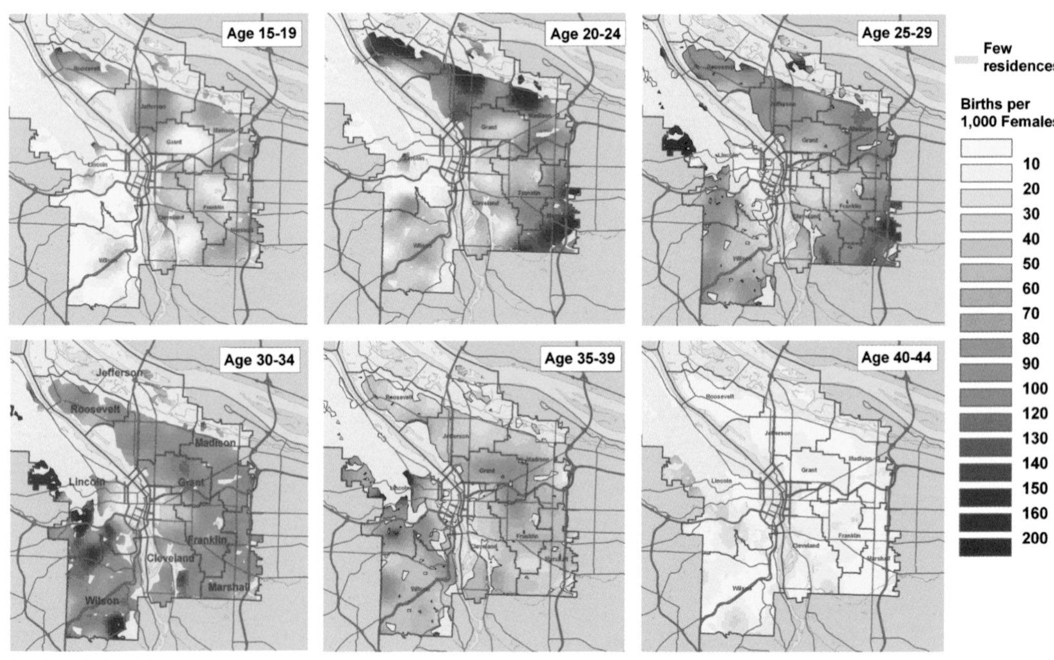

Age-Specific Birth Rates

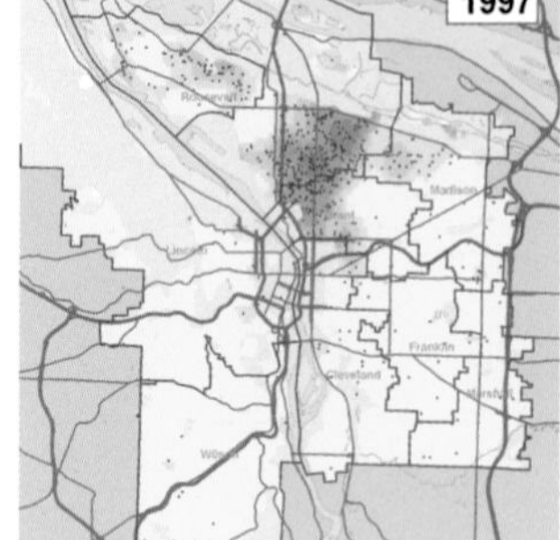

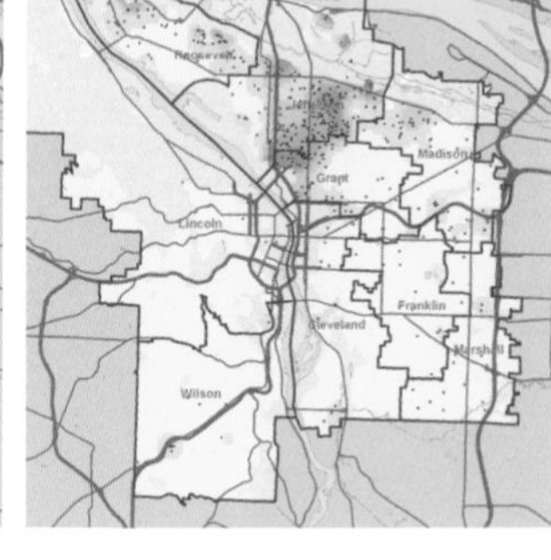

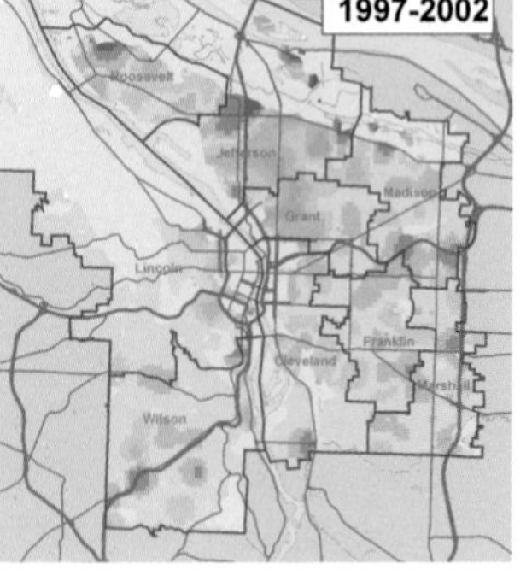

Few Residences
■ Black Births

Percent Births Black	% Change 1997 - 2002 Fewer
10	-40.0
20	-20.0
30	-10.0
40	-2.5
50	2.5
60	10.0
70	20.0
80	40.0
90	More

Births to Black Mothers

This project illustrates an application of spatial demographic analysis in support of school planning. The study of fertility is one of the key fields of demography, and fertility trends are one of the key determinants of school enrollment. The maps illustrate several standard demographic measures of fertility such as crude birthrates, general fertility rates, and age- and race-specific fertility rates. Also illustrated is the path from birth to kindergarten, as moderated by migration and school capture rates. The maps were created in ArcMap using the ArcGIS Spatial Analyst extension.

Courtesy of Portland State University.

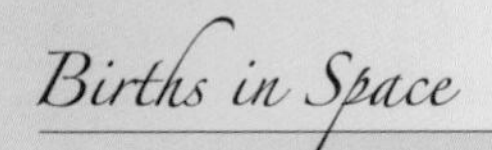

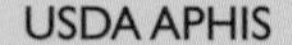

USDA APHIS
Raleigh, North Carolina, USA
By Yu Takeuchi

Contact

Yu Takeuchi
yu.takeuchi@aphis.usda.gov

Software

ArcGIS Desktop 9

Data Source(s)

USDA

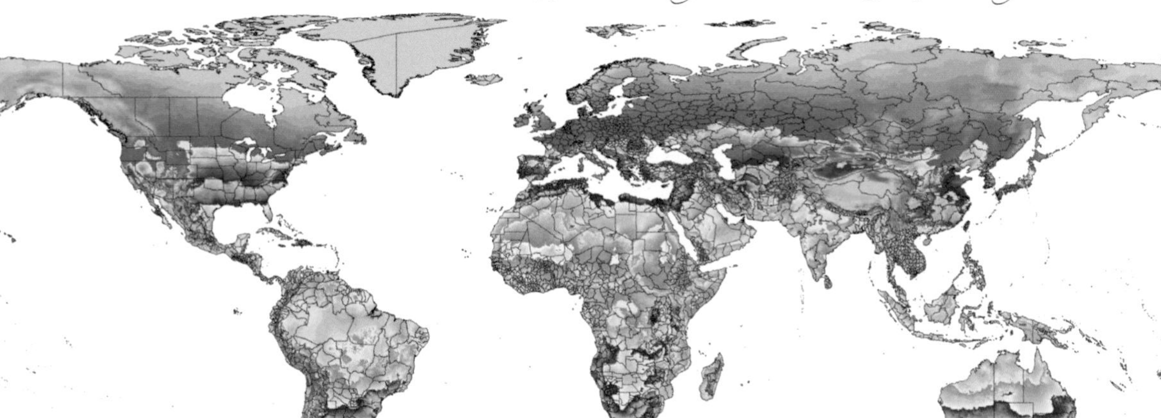

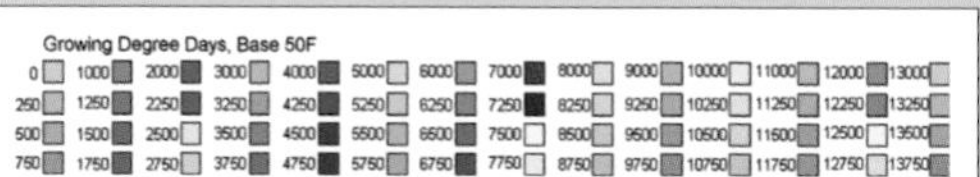

Potential Generations per Year for *Bactrocera invadens* in Africa

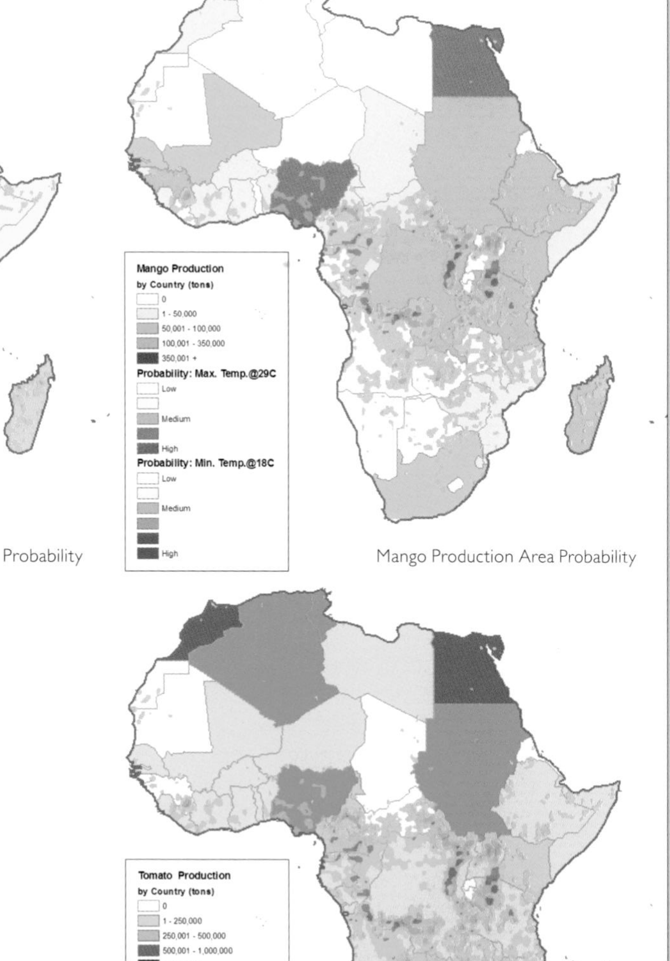

Citrus Production Area Probability

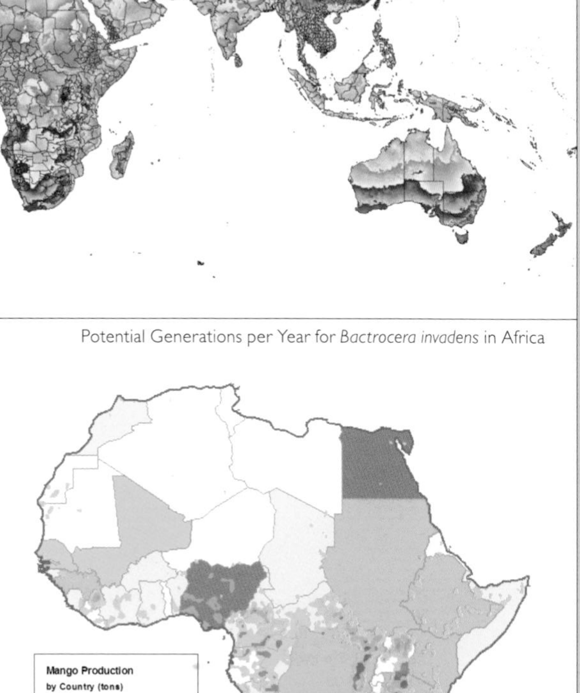

Mango Production Area Probability

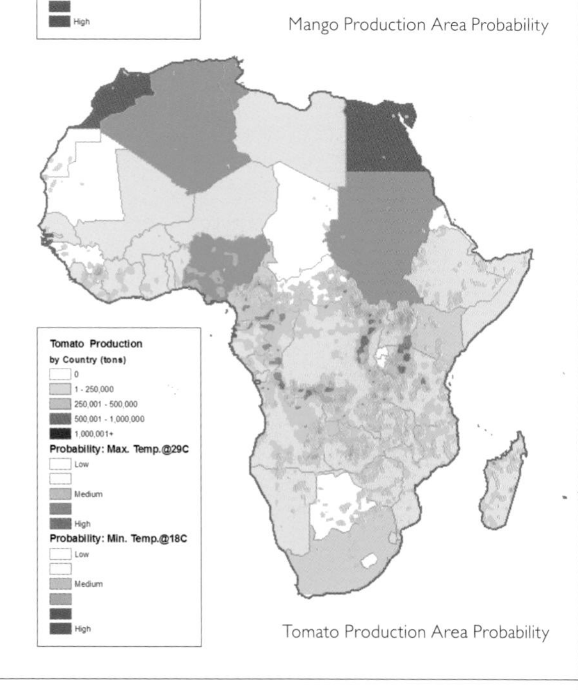

Tomato Production Area Probability

A joint venture between USDA Animal and Plant Health Inspection Service (APHIS), North Carolina State University (NCSU), and ZedX, Inc., the NCSU/APHIS Plant Pest Forecast System (NAPPFAST) is an Internet tool for plant pest modeling. NAPPFAST uses a graphical user interface to link meteorological and geographic databases with templates for biological modeling.

These maps illustrate forecasts of *Bactrocera invadens*, a new fruit fly that is rapidly spreading throughout central Africa. By using NAPPFAST, high-risk areas were determined based on optimum temperature for another fruit fly, *B. dorsalis*. These maps indicate probabilities of having maximum temperature and minimum temperature with host species production volume. Studies conducted thus far have determined the optimum temperature for fruit fly species is 29/18°C (maximum/minimum) on the basis of fecundity. Central Africa has the most suitable conditions for *Bactrocera* fruit fly development.

A generation map, created in ArcGIS, indicates the potential generations per year for *B. invadens*. Temperature is an important factor for the developmental rate of many organisms where a certain amount of heat is needed for development. By using degree days required for fruit flies and the NAPPFAST system, possible generation numbers were determined. A combination of spread forecast, pest biology, distribution, and importance of agricultural commodities and distribution of wild hosts are major elements in the determination of risk of *B. invadens*.

Courtesy of Yu Takeuchi, USDA APHIS.

Natural Resources Conservation Service,
U.S. Department of Agriculture
Morgantown, West Virginia, USA
By Sharon Waltman, Melissa Marinaro, and James T. Carrington

Contact

Melissa Marinaro
melissa.marinaro@wv.usda.gov

Software

ArcGIS Desktop

Printer

HP Designjet 5500ps

Data Source(s)

SSURGO, NEDS, NASIS, TIGER, National Atlas

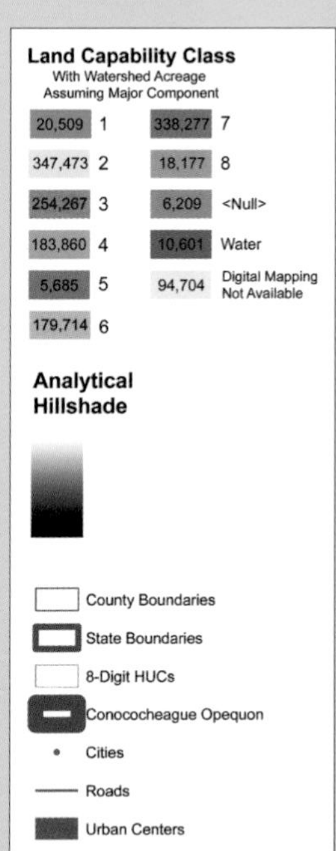

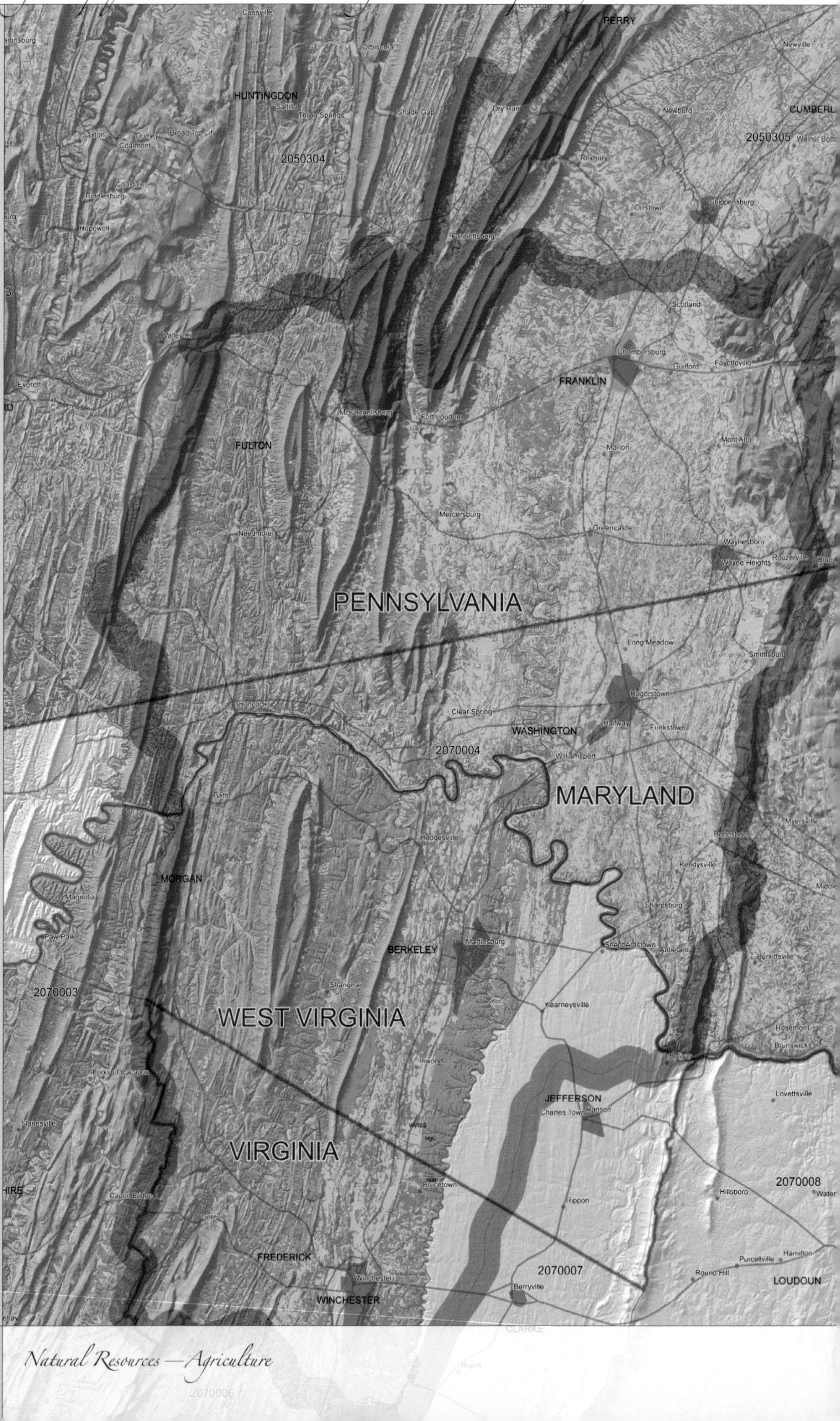

Both maps are from the Detailed Soil Survey Atlas collection of interpretive maps and illustrate application of detailed soil survey information obtained from the Soil Survey Geographic Database (SSURGO) for natural resource planning. These maps are important for making land-use management decisions regarding the soil resource. The Conococheague-Opequon is a subwatershed of the Chesapeake Bay Watershed and covers portions of Maryland, Pennsylvania, Virginia, and West Virginia.

Land capability classification is a system of grouping soils primarily on the basis of the soil's enduring ability to produce common cultivated crops and pasture plants without deteriorating. Class codes 1 to 8 are used here to represent nonirrigated land capability. Definitions range from Class 1 soils, which have slight limitations that restrict their use, to Class 8, whose limitations preclude their use for commercial plant production, restricting them to recreation, wildlife, water supply, or aesthetic purposes.

Courtesy of Natural Resources Conservation Service, U.S. Department of Agriculture.

Natural Resources Conservation Service, U.S. Department of Agriculture
Morgantown, West Virginia, USA
By Sharon Waltman, Melissa Marinaro, and James T. Carrington

Contact
Melissa Marinaro
melissa.marinaro@wv.usda.gov

Software
ArcGIS Desktop

Printer
HP Designjet 5500ps

Data Source(s)
SSURGO, NEDS, NASIS, TIGER, National Atlas

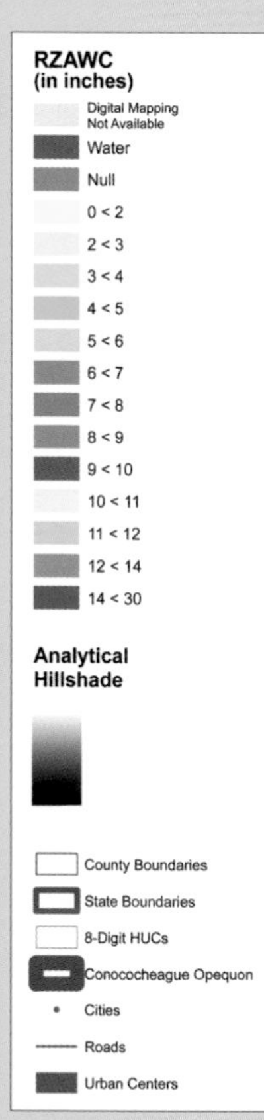

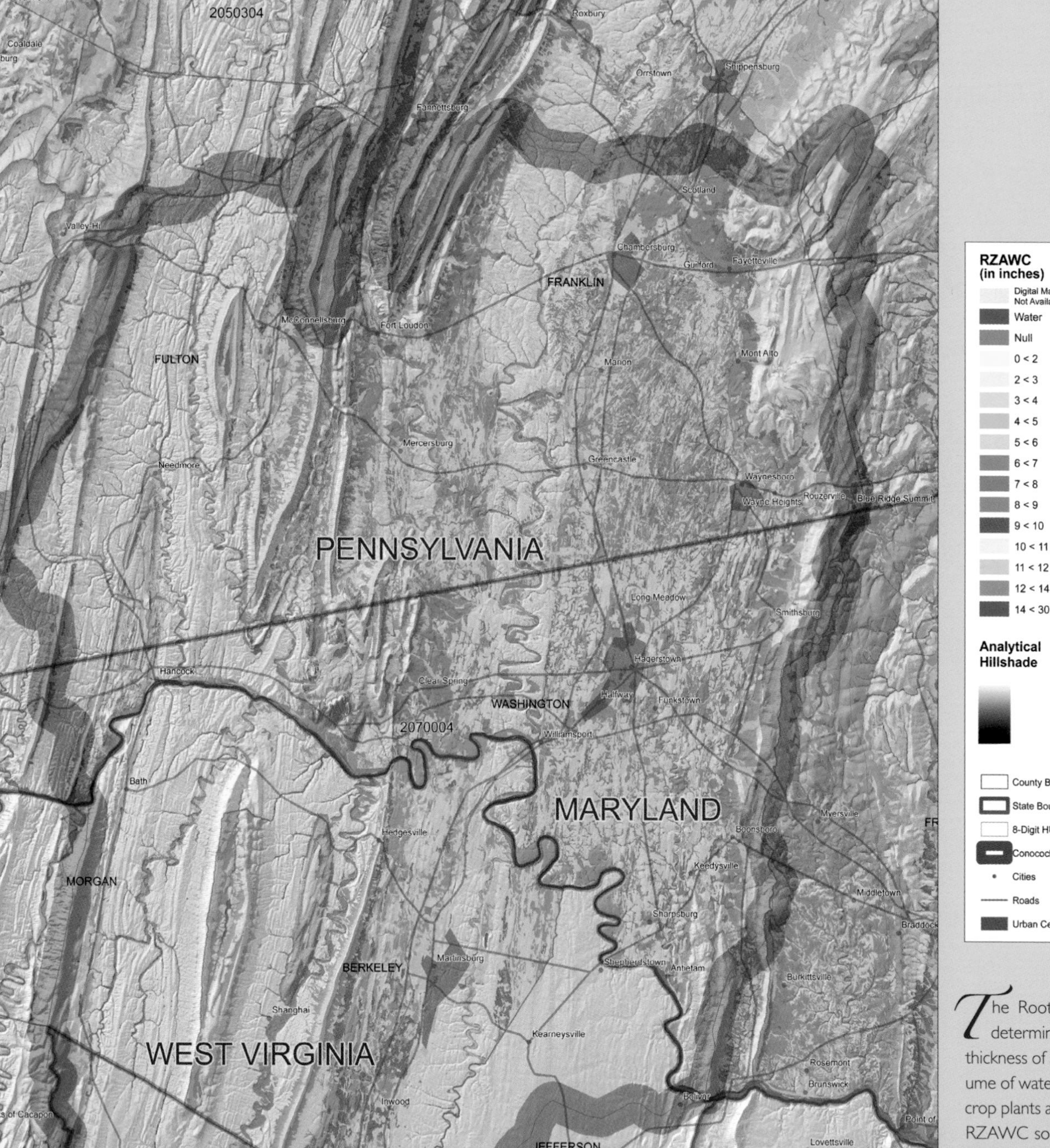

The Root Zone Available Water Capacity (RZAWC) is determined by multiplying available water capacity by the thickness of all layers in the root zone. It approximates the volume of water that is held in the root zone that can be used by crop plants and reflects important physical soil properties. The RZAWC soil is calculated from the surface to the beginning of the first root-restrictive soil layer, such as bedrock or a very dense layer, or to a depth of 150 centimeters. The importance of RZAWC varies geographically because the capacity to hold water during key parts of the growing season is more critical in some climates than in others. Available water in the surface layer is critical to establish plants, but the amount of available water stored throughout the root zone usually determines the most productive soils for the entire growing season.

Courtesy of Natural Resources Conservation Service, U.S. Department of Agriculture.

Natural Resources Conservation Service
Beltsville, Maryland, USA
By Dean Oman and Robert Kellogg

Contact
Dean Oman and Robert Kellogg
dean.oman@wdc.usda.gov

Software
ArcGIS Desktop 9

Printer
HP Designjet 5000ps

Data Source(s)
1997 National Resources Inventory (NRI USDA-NRCS), 1990–1995 Cropping Practices Surveys (USDA ERS 2000), 1991–1993 Area Studies Survey (USDA-NASS), Crop Residue Management Survey (CRMS), Conservation tillage data (CTIC-2001), EPIC climatic dataset

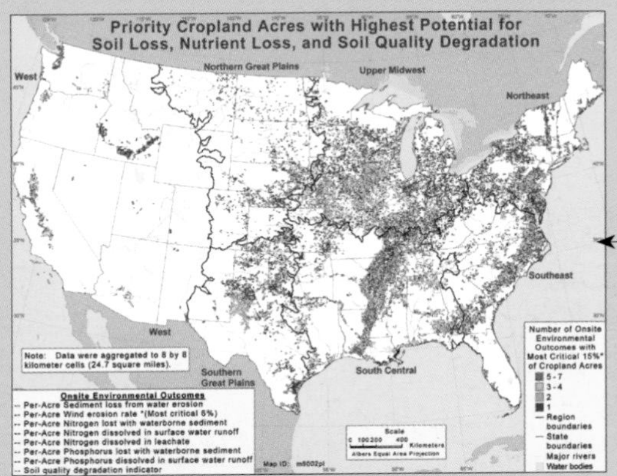

Top 15 Percent of Acres from Each Map Combined

The National Resources Inventory (NRI) provided the analytical framework for a simulation modeling approach to identify cropland acres with the greatest potential for soil loss, nutrient loss, and soil quality degradation. The physical process model EPIC (Erosion-Productivity Impact Calculator) was used to estimate surface-water runoff, percolation, wind erosion, sediment loss, nutrient loss, and changes in soil organic carbon for each NRI cropland sample point included. More than 750,000 EPIC model runs were conducted. Model results were estimated for 15 crops representing approximately 298 million acres, or 79 percent, of U.S. cropland, exclusive of acres enrolled in the Conservation Reserve Program.

The map above shows priority cropland acres for conservation program implementation based upon estimates illustrated in the eight maps to the right, which show potential losses of soils and nutrients from farm fields and the potential for soil quality degradation. Priority acres—those most in need of conservation treatment—are critical for one or more of the eight on-site environmental outcomes. For each outcome, critical acres were identified as those with the highest loss estimates (or lowest soil quality rating) in the country, generally representing those in the top 15 percent nationally.

For more details see "Model Simulation of Soil Loss, Nutrient Loss, and Change in Soil Organic Carbon Associated with Crop Production" by Potter, Andrews, Atwood, Kellogg, Lemunyon, Norfleet, and Oman. June 2006. U.S. Department of Agriculture, Natural Resources Conservation Service. www.nrcs.usda.gov/Technical/nri/ceap/croplandreport

Courtesy of Natural Resources Conservation Service.

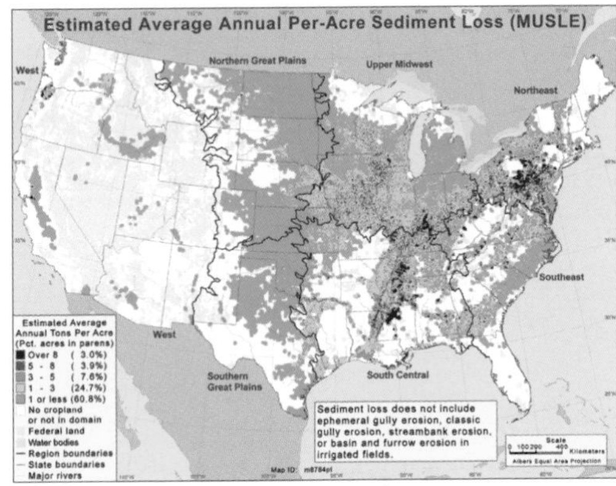

Sediment Loss from Water Erosion

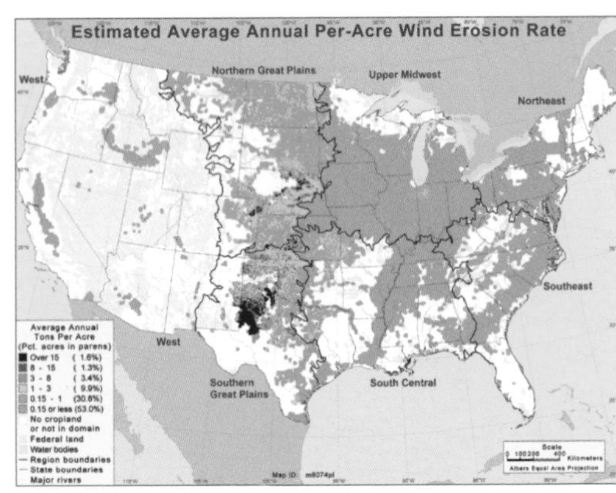

Wind Erosion Rate

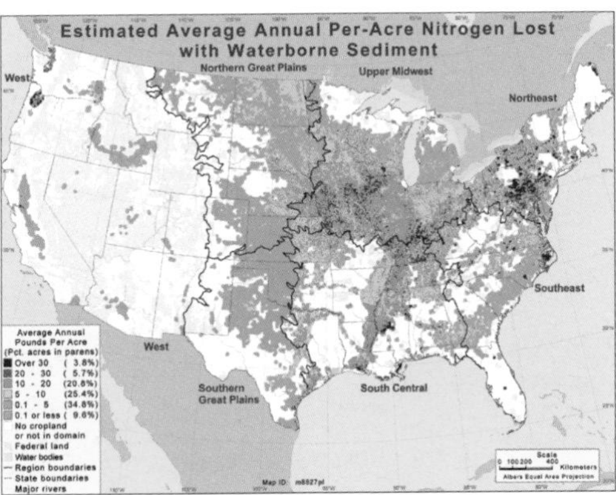

Nitrogen Loss with Waterborne Sediment

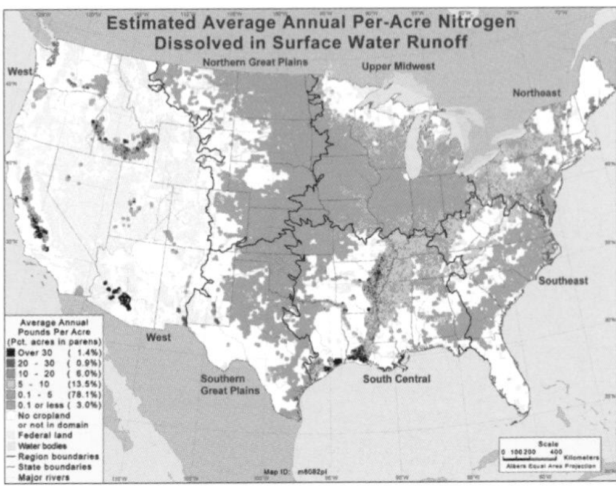

Nitrogen Dissolved in Surface Water Runoff

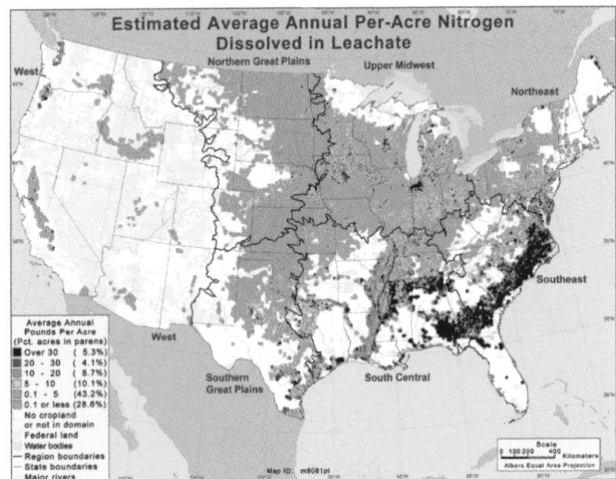

Nitrogen Dissolved in Leachate

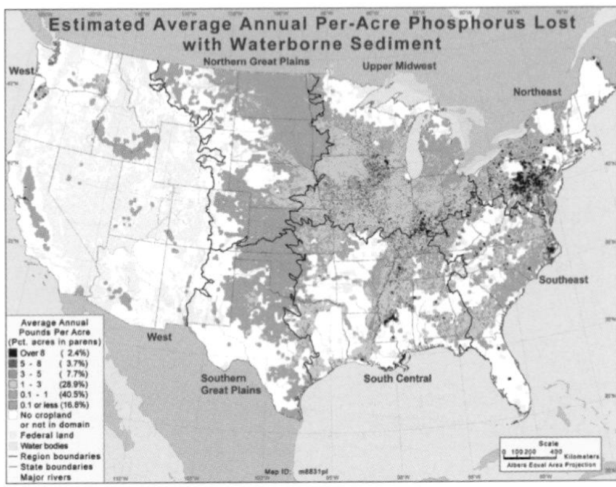

Phosphorus Loss with Waterborne Sediment

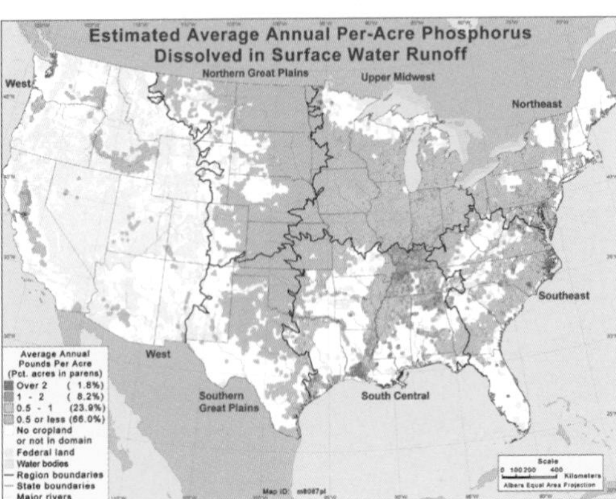

Phosphorus Loss in Surface Water Runoff

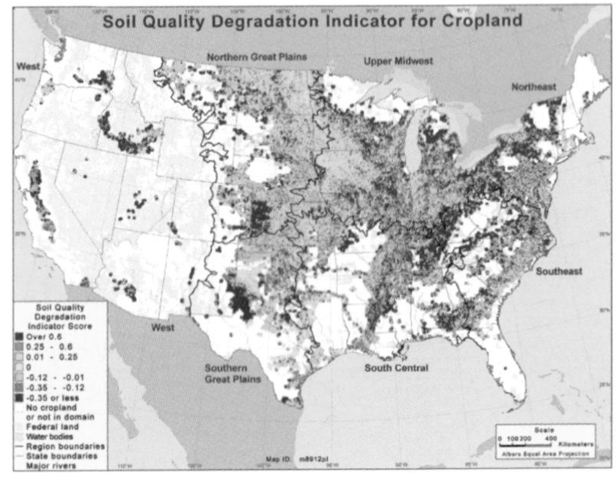

Soil Quality Degradation

Timberline Natural Resource Group, Ltd.
Vancouver, British Columbia, Canada
By Marcel Morin

Whitefeather Forest Management Corporation
Pikangikum, Ontario, Canada

Contact
Marcel Morin
marcel.morin@timberline.ca

Software
ArcInfo, Adobe Illustrator, Adobe Photoshop

Printer
HP Designjet 5500

Data Source(s)
Digital Forest Resource Inventory, indigenous knowledge (elders of Pikangikum)

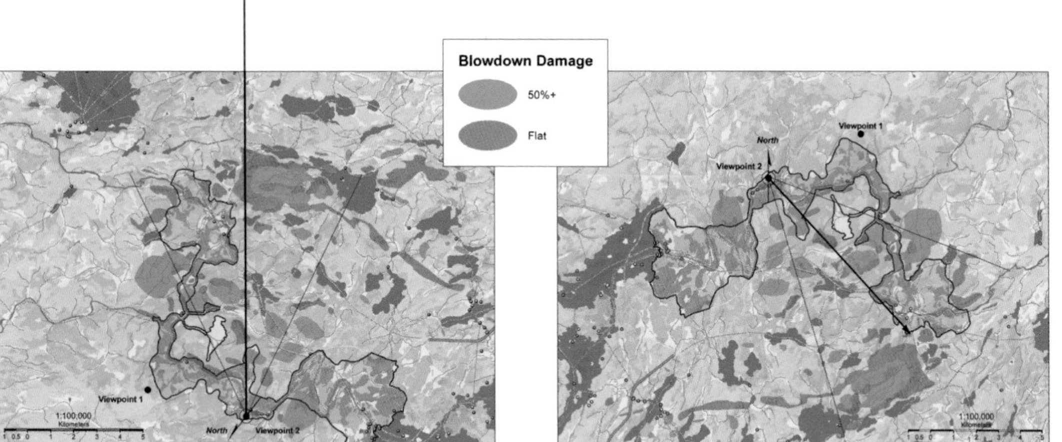

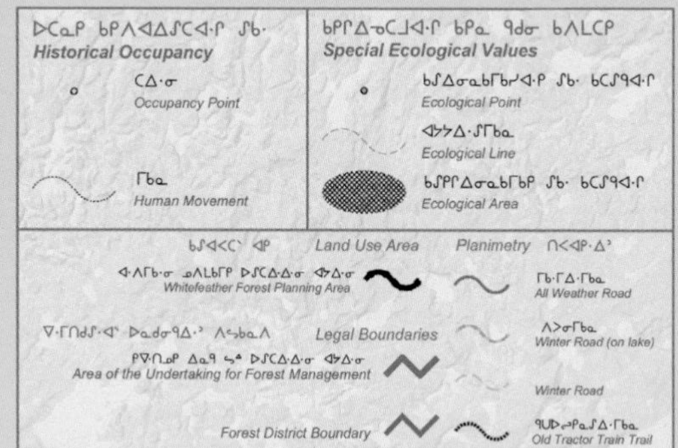

In 2004, a violent storm raced across the Whitefeather Forest within the ancestral lands of Pikangikum First Nation in northwestern Ontario. The storm included a severe tornado that destroyed approximately 15,000 hectares of valuable northern boreal forest timber in the Whitefeather Forest. As part of the planning process for the Whitefeather Forest Initiative, the Whitefeather Forest Management Corporation directed the preparation of a computer viewscape analysis of how the blowdown would be seen from various vantage points along the Berens River and what the zone would look like if forestry operations had occurred in the area affected by the tornado.

Area Dedications Map

Timberline Natural Resource Group, Ltd.
Vancouver, British Columbia, Canada
By Marcel Morin

Whitefeather Forest Management Corporation
Pikangikum, Ontario, Canada

Contact
Marcel Morin
marcel.morin@timberline.ca

Software
ArcInfo, Adobe Illustrator, Adobe Photoshop

Printer
HP Designjet 5500

Data Source(s)
Digital Forest Resource Inventory,
indigenous knowledge (elders of Pikangikum)

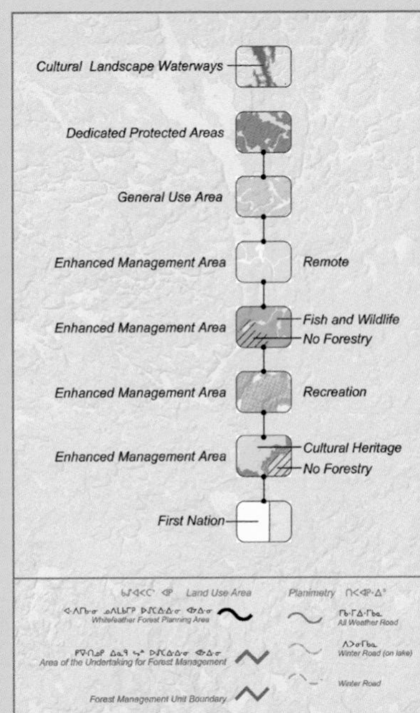

In 2006, the Ontario Minister of Natural Resources traveled to Pikangikum First Nation in northwestern Ontario to participate in a celebration and feast for the joint approval of the landmark Whitefeather Forest and Adjacent Areas Land Use Strategy as official policy for Ontario and the Pikangikum people. A community-based land-use planning approach to develop the strategy was led by the Pikangikum and involved a strong, cooperative, consensus-based decision-making partnership with Ontario. This approach was used to identify where new economic opportunities including forestry and tourism can be developed in the 1.3 million-hectare planning area. Planning was based on the best of indigenous knowledge and western science. The composite map in the Land Use Strategy identifies areas for new economic opportunities where the Pikangikum will take a leading role while ensuring that "all living ones" are sustained through a "keeping the land" approach.

McInnes Lake
Flanagan
Barton Lake
Dolphin Lake
Throat River
Beekahncheekahmeeng
Pikangikum
Pikangikum Lake
Berens River
Berens Lake
Silcox Lake
Sahkeesahkahteekoh
weesahkaheegahn
Pringle Lake
Ohshkahtahkah weesuhkaheekahneeng
Meesheekeenaypeekohseepee
Nungesser Lake North
Pringle Lake
Kirkness Lake
Kahnahmaykoosayseekahk
Valhalla - Trough Lake

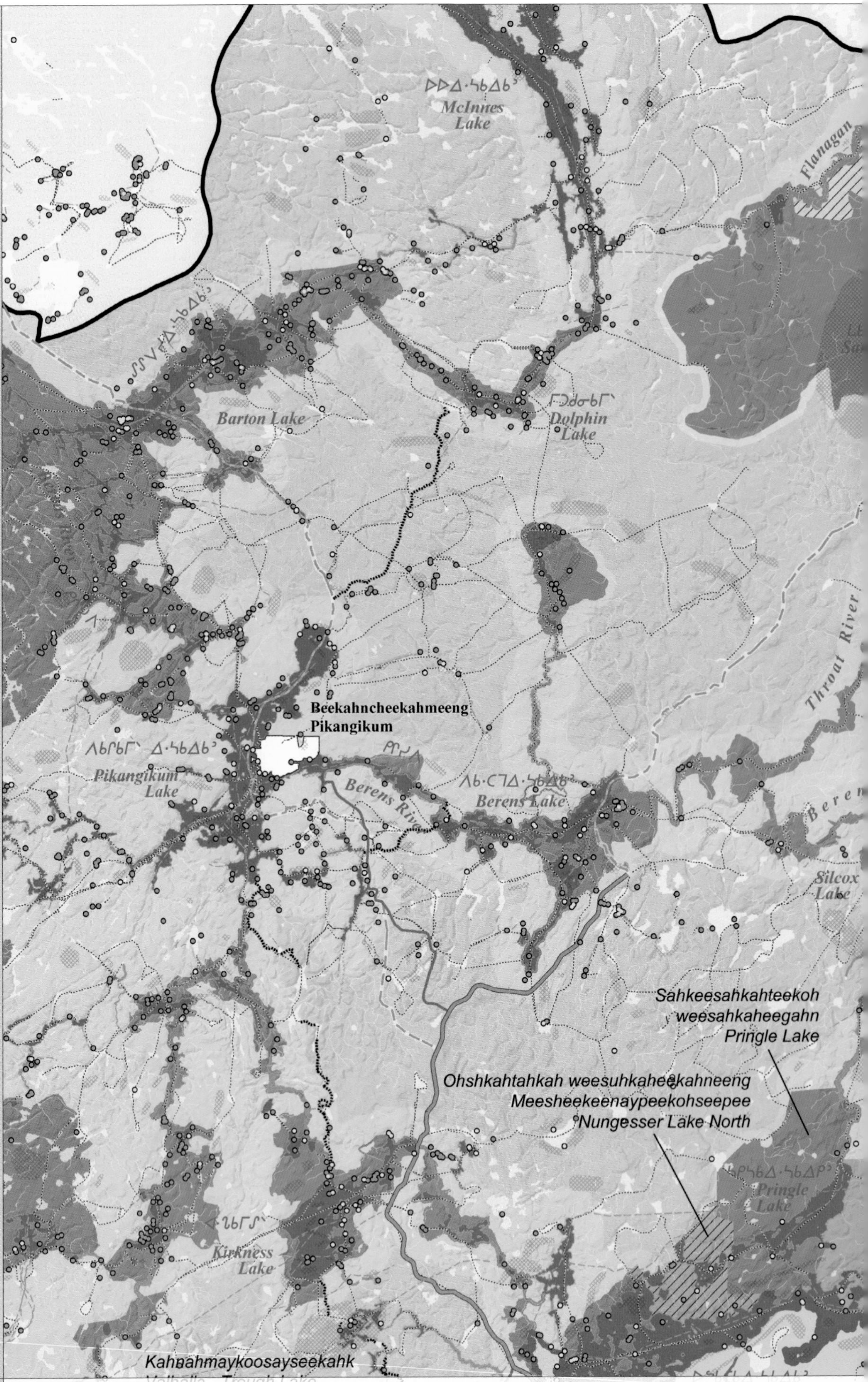

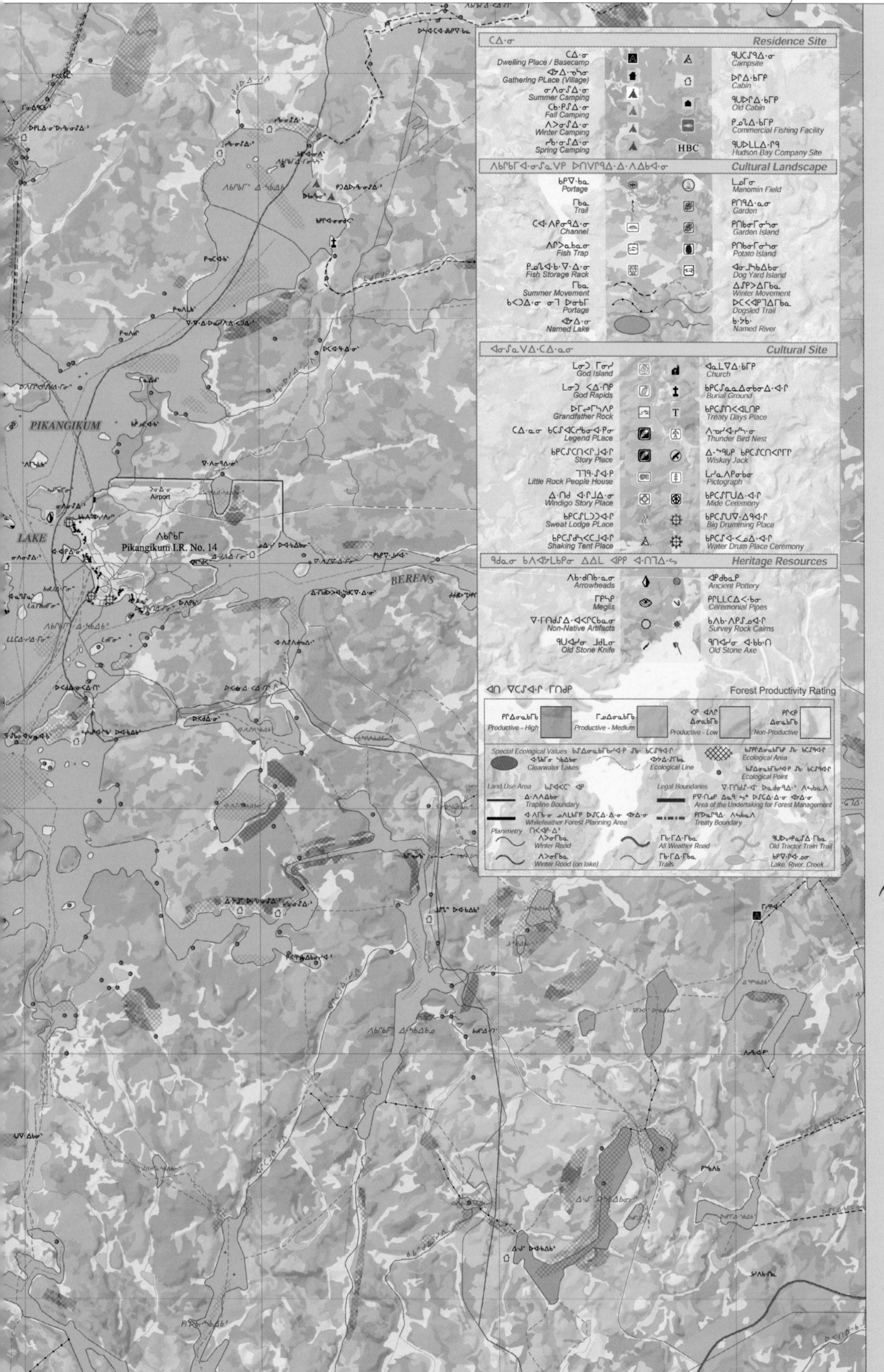

Timberline Natural Resource Group, Ltd.

Vancouver, British Columbia, Canada
By Marcel Morin

Whitefeather Forest Management Corporation

Pikangikum, Ontario, Canada

Contact

Marcel Morin
marcel.morin@timberline.ca

Software

ArcInfo, Adobe Illustrator, Adobe Photoshop

Printer

HP Designjet 5500

Data Source(s)

Digital Forest Resource Inventory, indigenous knowledge (elders of Pikangikum)

The Whitefeather Forest is an indigenous cultural landscape of the Pikangikum people. The Pikangikum have protected and enhanced the biodiversity of the landscape and nurtured the abundance of its diverse resources. They have achieved this through customary indigenous resource stewardship practices and management tools supported by a rich indigenous knowledge tradition. From its vast tracts of jack pine and wild rice fields planted by the Pikangikum people, to rich muskrat marshes and abundant wildlife species, the Whitefeather Forest cultural landscape is of international ecological significance.

The ecological richness of the Whitefeather Forest landscape is complemented by a cultural heritage that includes features such as pictographs, campgrounds, portages, and canoe channels. These enhance the numerous pristine waterways that flow through the forest. It is the intention of the Pikangikum First Nation in the Whitefeather Forest Initiative to provide economic opportunities for community members while preserving the rich ecological and cultural heritage of the forests.

Forest Land Ownership in the Conterminous United States

USDA Forest Service
St. Paul, Minnesota, USA
By Mark D. Nelson and Greg C. Liknes

Contact
Mark D. Nelson
mdnelson@fs.fed.us

Software
ArcGIS Desktop 9.1 and ArcGIS Spatial Analyst

Data Source(s)
USFS, USGS, CBI, EPA, ESRI

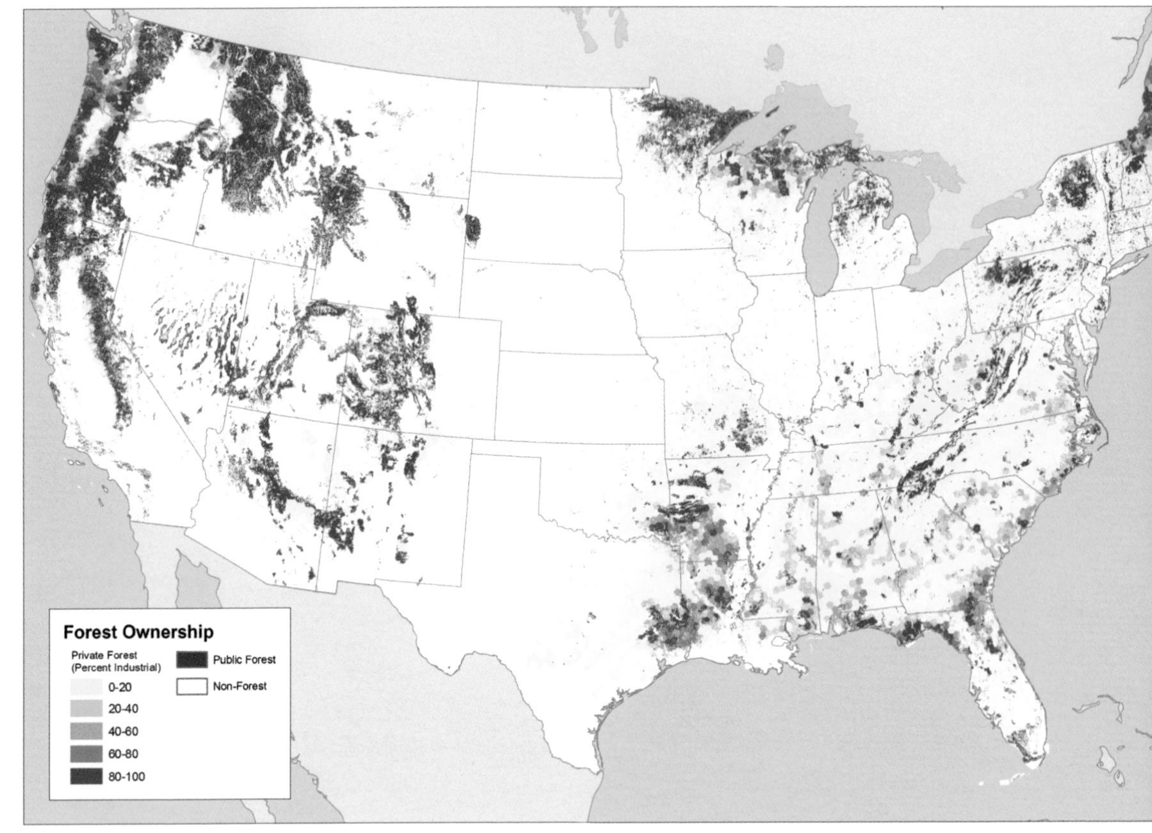

Patterns of public and private forestland ownership vary across the United States. For example, two-thirds of western forestland is publicly owned, mostly by federal agencies such as the U.S. Forest Service (USFS), Bureau of Land Management, and National Park Service. However, more than 80 percent of eastern forestland is privately owned. Private forestland is further differentiated by industrial ownership classes. Private industrial forest is land owned by a company or corporation that operates a primary wood-using plant, such as a sawmill or paper mill. The heaviest concentrations of private industrial forestland are in the Northeast, South, and Pacific Northwest.

The pie charts were designed such that the colors correspond to the map categories. Public ownership appears in blue, while private forest is differentiated by industrial (red) versus nonindustrial (yellow) ownership. The charts summarize the percentage of forestland by ownership category. The size (area) of each pie chart is proportional to the total forest area in each region. Comparing these charts, it becomes apparent how drastic the difference in public versus private ownership is from east to west.

The percentage of private industrial forestland was estimated from Forest and Rangeland Renewable Resources Planning Act (RPA) data, derived primarily from the USFS Forest Inventory and Analysis (FIA) program. The data was summarized over a hexagon sampling array developed by the U.S. Environmental Protection Agency Environmental Monitoring and Analysis Program (EMAP). The forestland spatial pattern was derived from the U.S. Geological Survey National Land Cover Dataset (NLCD) and was classified as public or private ownership according to the Protected Areas Database of the Conservation Biology Institute. Each forestland pixel classified as private was then attributed with the percentage industrial ownership value associated with the EMAP hexagon in which the pixel was located. Political boundaries were derived from ESRI Data & Maps.

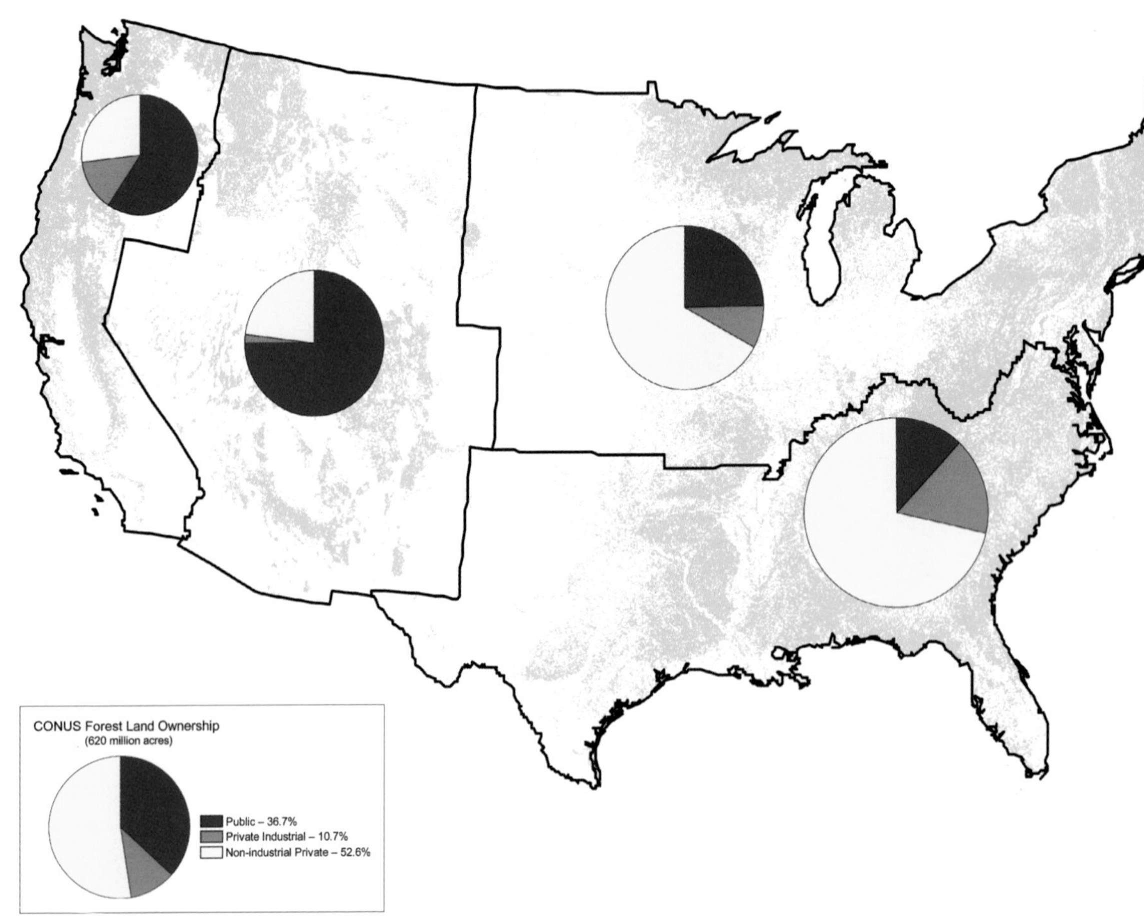

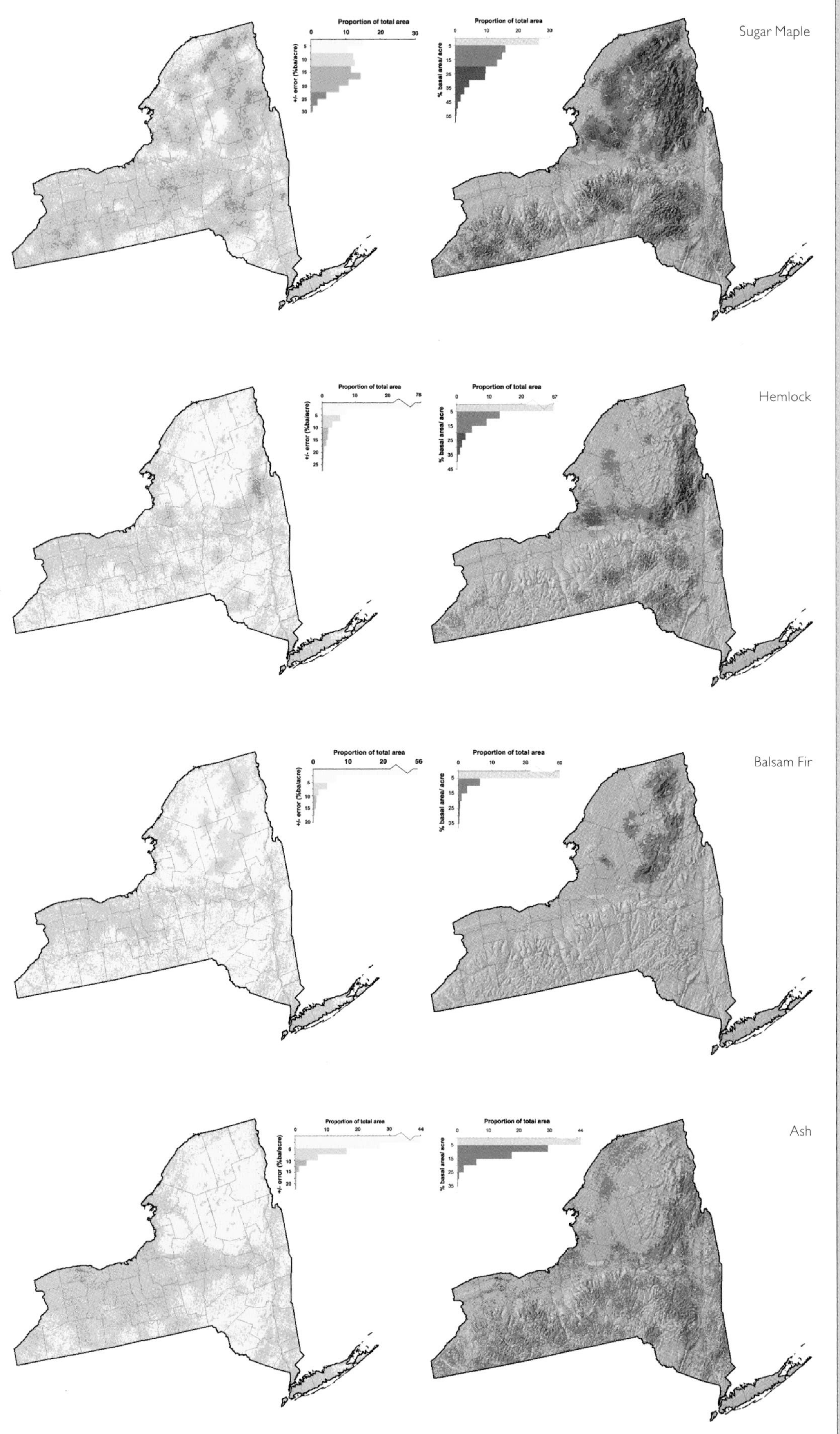

USDA Forest Service
Troy, New York, USA
Rachel Riemann, Sierra Murdoch, and Andy Lister

Contact
Rachel Riemann
rriemann@fs.fed.us

Software
ArcGIS Desktop 9.1

Data Source(s)
USDA Forest Service, Forest Inventory and Analysis periodic data, National Atlas

The goal of this project was to create maps of forest characteristics—in this case, distribution and relative importance of different tree species—from data collected by the U.S. Forest Service Forest Inventory and Analysis (USFS FIA) program. The maps were generated from USFS FIA plot data from the last periodic inventory in each state (1993 for New York). They were created by stochastic simulation using Geostatistical Software Library (GSLIB), while modeling was done using a two-kilometer grid.

Each map provides an estimate of relative total tree-stem cross-sectional area per acre of that species (percentage basal area per acre). Since any modeled estimate inevitably includes some level of uncertainty, each map is accompanied by a mapped estimate of uncertainty. The legends show graphs of the proportion of forestland belonging to both the percentage basal area/acre classes and the uncertainty classes.

These species maps also exist as posters that include 12 species for each state. Four posters currently exist: one each for New York, Vermont, New Hampshire, and southern New England (Massachusetts, Connecticut, and Rhode Island). Maps with the most recent FIA data are currently being developed. The goal is to generate spatial datasets for the 24-state area covered by the Northern Research Station (from the Dakotas to Maine).

Courtesy of USDA Forest Service.

Los Alamos National Laboratory, EES-9
Los Alamos, New Mexico, USA
By Richard E. Kelley

Contact
Richard E. Kelley
rekelley@lanl.gov

Software
ArcGIS Desktop 9.1, ArcGIS Spatial Analyst, ArcGIS 3D Analyst

Printer
HP Designjet 5500

Data Source(s)
USGS and other public data

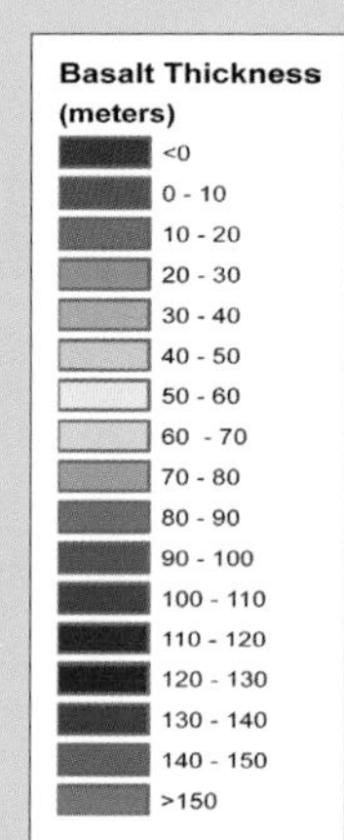

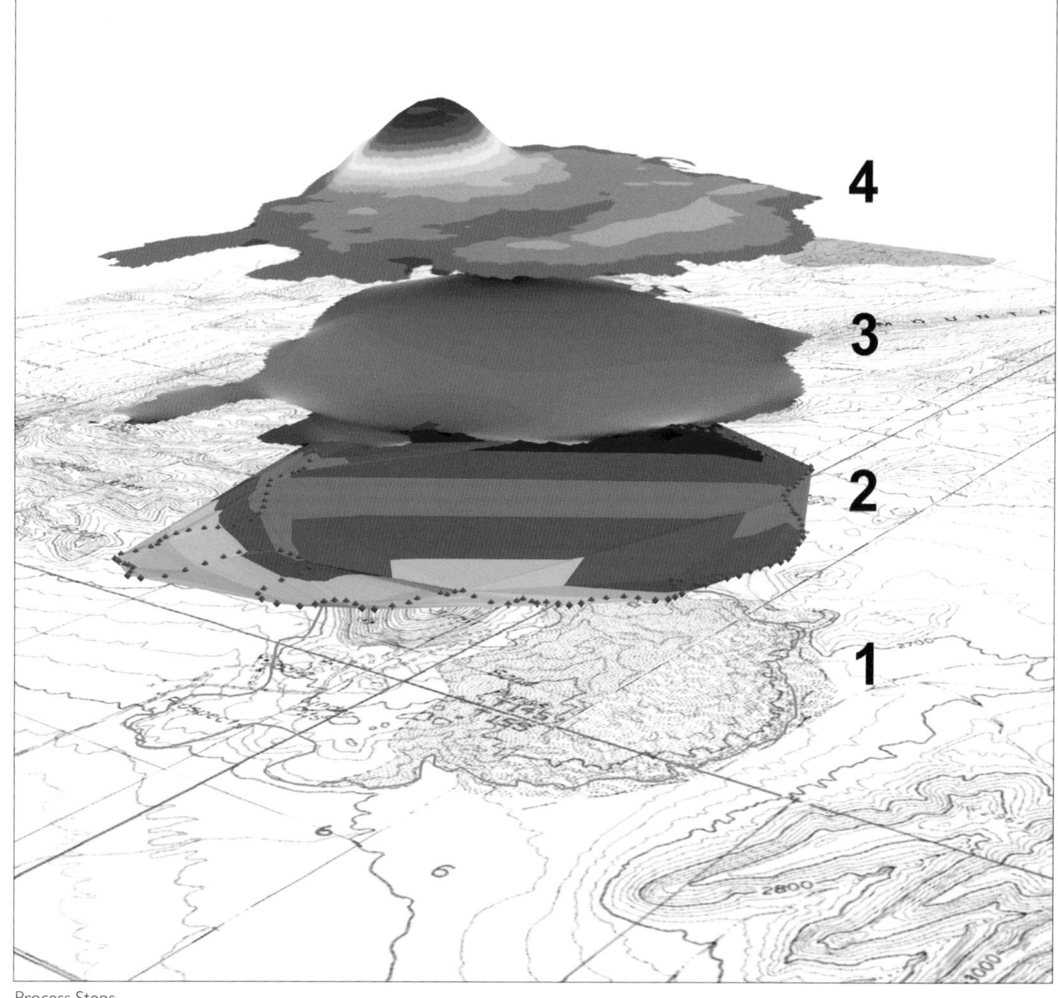

Process Steps

As part of the Yucca Mountain Probabilistic Volcanic Hazard Analysis (PVHA) Project, small Quarternary (0.3–1.0 Ma) volcanoes in the vicinity of the proposed Yucca Mountain repository were studied. Included were calculations of the erupted magma volume for each volcano. ArcGIS Desktop with ArcGIS 3D Analyst was used to perform this task. The steps used in making the calculations and 3D graphic representations of the volcanoes are displayed in this map.

Process Steps

1. The geological extent of the basalt cone/flow is defined from published geologic maps or air photo data and used as a mask to extract a "top of basalt" raster from 10-meter digital elevation model (DEM) data.
2. The outline of the basalt extent is converted to three-dimensional points. From this, a triangulated irregular network (TIN) is created, representing an approximation of the topographic surface that was present prior to eruption.
3. The TIN is then converted to a raster surface, representing the base of the erupted basalt.
4. The base elevation raster is then subtracted from the "top of basalt" raster, resulting in a "thickness" raster. The volume of basalt is calculated with the "Area and Volume Statistics" tool in ArcGIS 3D Analyst.

Courtesy of Los Alamos National Laboratory, EES-9.

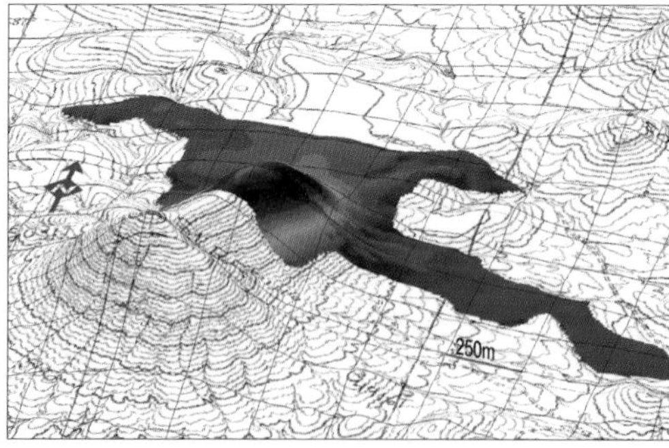

Hidden Cone

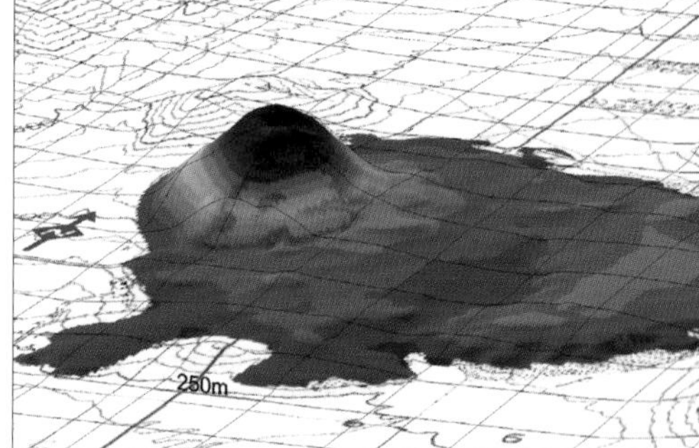

Latrop Wells

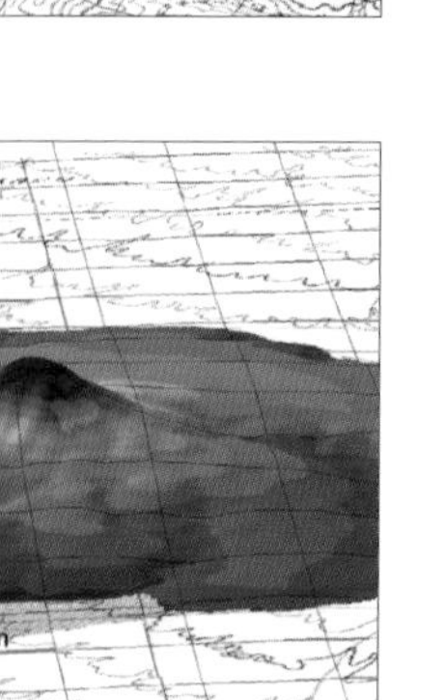

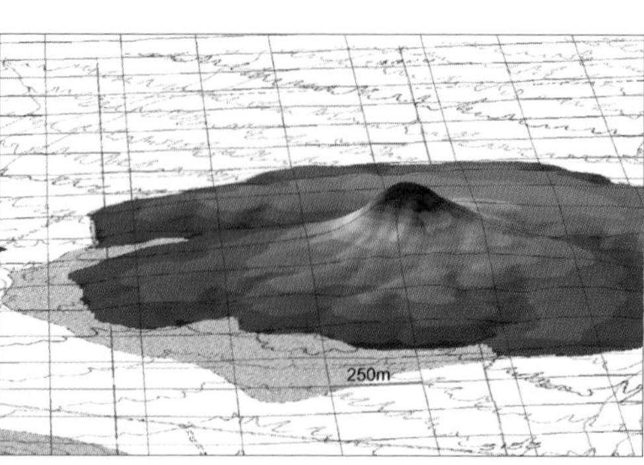

Black Cone

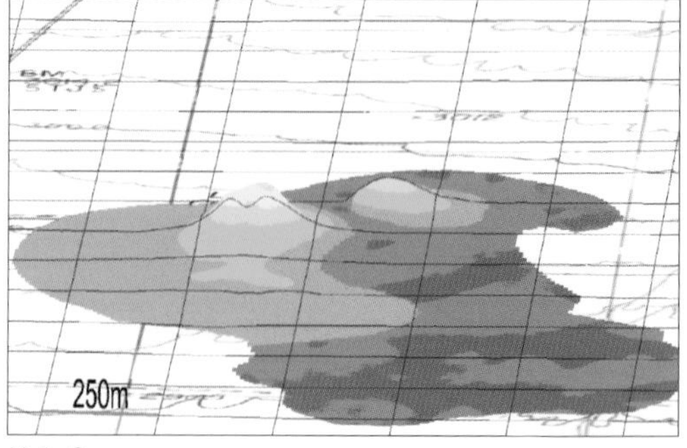

Little Cones

Ludwig Maximilians University, Department of Earth and Environmental Science
Munich, Germany
By L. Dolezalek, K. Scharrer, and U. Muenzer

Contact
Lorenz Dolezalek
LDolezalek@munichre.com

Software
ArcView, ArcGIS 3D Analyst

Printer
HP Designjet 1050

Data Source(s)
Icelandic Meteorological Office

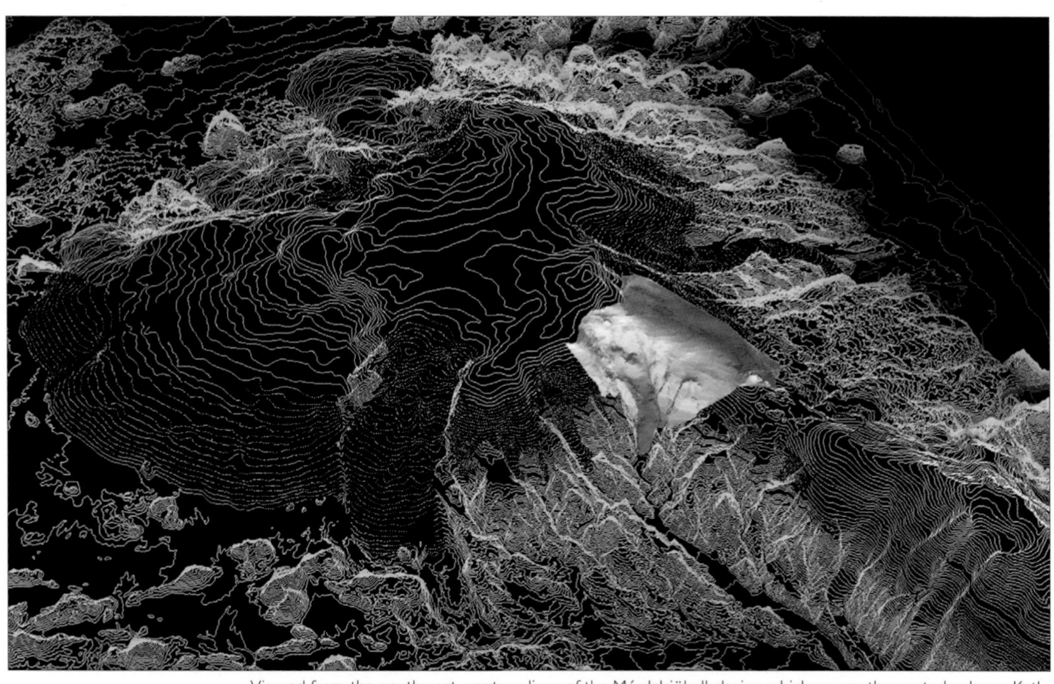

Viewed from the northwest, contour lines of the Mýrdalsjökull glacier, which covers the central volcano Katla.

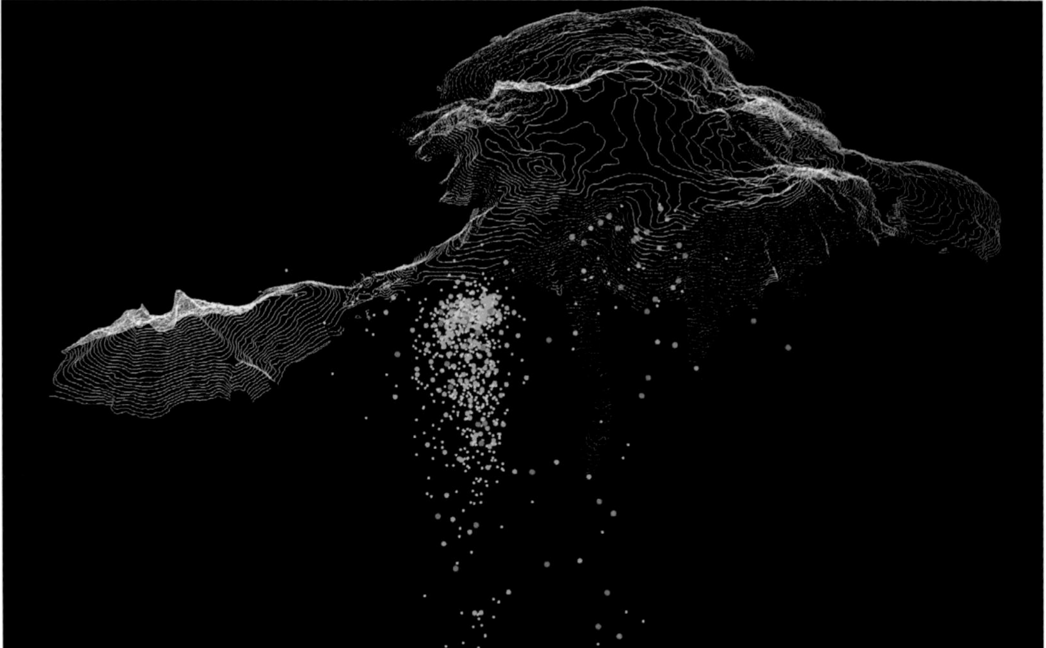

Viewed from the southwest, earthquake hypocenters underneath the Mýrdalsjökull Glacier.

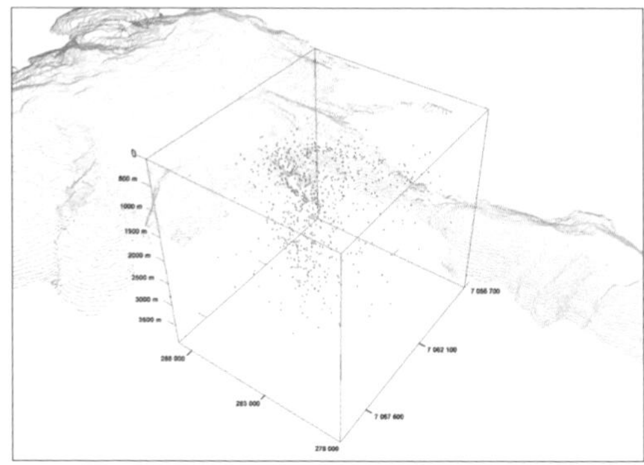

Viewed from the northwest, earthquake hypocenters underneath the Mýrdalsjökull Glacier.

Mýrdalsjökull Glacier, Landsat, copyright NASA

The Katla central volcano is among the most active and dangerous subglacial volcanoes in Iceland. The caldera is covered by Mýrdalsjökull, the fourth largest Icelandic glacier (586 km²). The average eruption frequency of Katla is twice per century, with the last major eruption dating back to 1918. The maximum melt water discharge of the torrent accompanying this eruption has been estimated at 300,000 cubic meters per second.

The objective of the ongoing project Hazard Assessment and Prediction of Icelandic Volcanoes is the establishment of a largely operational monitoring service to be employed for risk assessment, early warning, and damage mitigation. To consider as many precursors of an imminent subglacial eruption as possible, a multisource geographic information system has been developed.

Indicated with contour lines, the maps show the subglacial volcano Katla covered by the Mýrdalsjökull glacier. Underneath the surface, reddish dots symbolize earthquake hypocenters. The distribution of the shown earthquake swarm marks a potential eruption site. Regarding the eruption cycle and the increase in seismic activity since 2000, a fresh outbreak releasing huge glacial torrents is expected in the near future.

Courtesy of Ludwig Maximilians University, Department of Earth and Environmental Science.

Geology of British Columbia

Clover Point Cartographics, Ltd.
Victoria, British Columbia, Canada
By Clover Point Cartographics, Ltd.

Contact
Mike Shasko
shasko.m@cloverpoint.com

Software
ArcInfo 9.1

Printer
HP Designjet 1055

Data Source(s)
Client data

The Geology of British Columbia map is a series of two maps and one detailed legend that depicts the geology of the province in full color at a 1:1,000,000 scale. It reflects the results of numerous field surveys by British Columbia Geological Survey geologists and their colleagues from federal government and universities. The map is a critical tool that can be used by a wide variety of users, including prospectors and mineral exploration companies, for identifying areas to search for new mines; geologists, for assessing concerns about local natural hazards; and teachers, for explaining the geology in their local area. Project partners include the British Columbia Ministry of Energy, Mines and Petroleum Resources, and Clover Point Cartographics, Ltd.

Courtesy of Clover Point Cartographics, Ltd.

Seafloor Morphology and Selected Features Between the Coastline and the Santa Lucia Escarpment, Central California

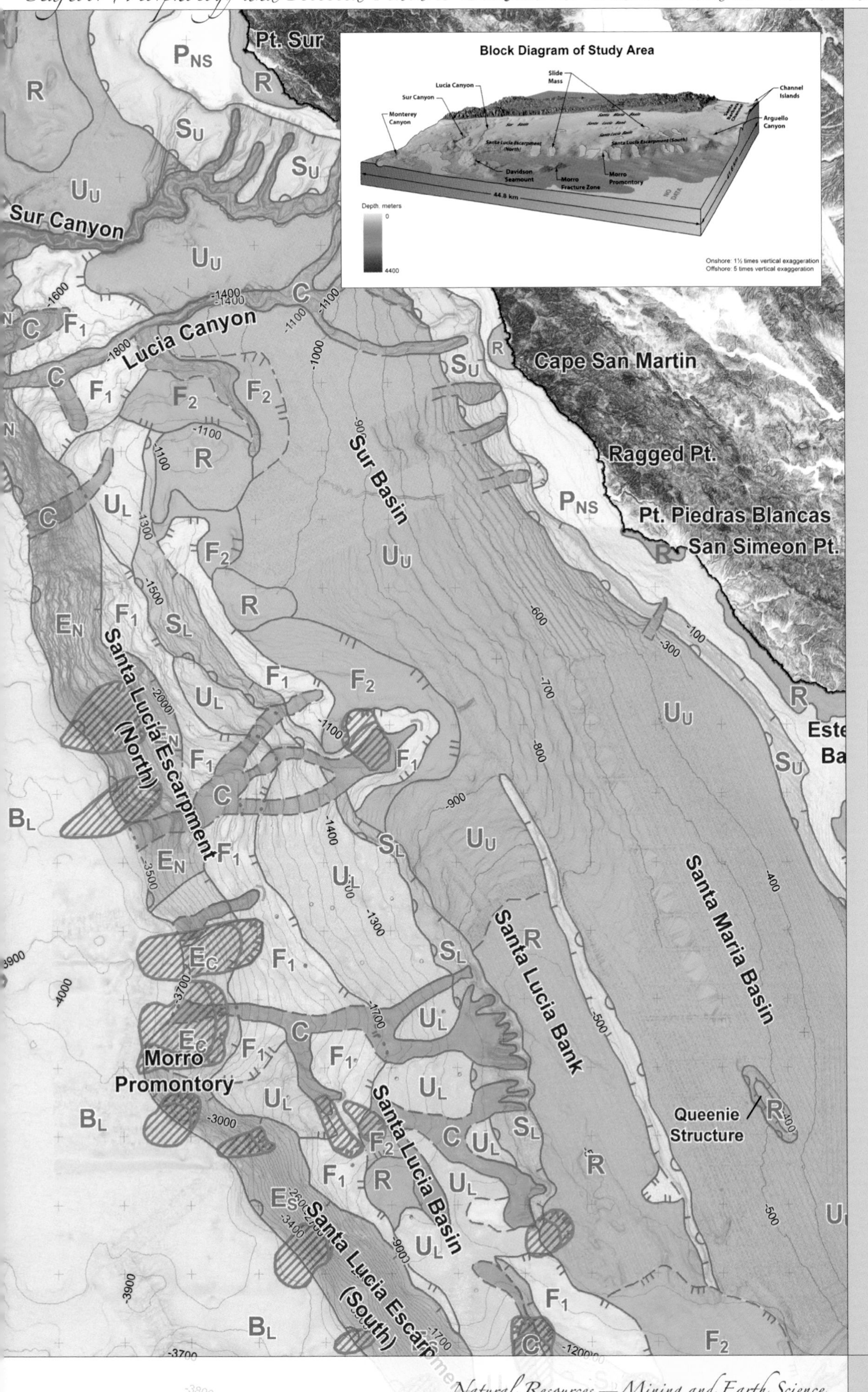

Geomatrix Consultants
Oakland, California, USA
By David O'Shea

Contact
David O'Shea
doshea@geomatrix.com

Software
ArcGIS Desktop 9, ArcGIS 3D Analyst, AutoCAD, Adobe Illustrator, Adobe Photoshop

Printer
HP Designjet 1055cm

Data Source(s)
USGS, NOS/NOAA multibeam bathymetry, NGDC/NOAA coastal relief DEM, MBARI

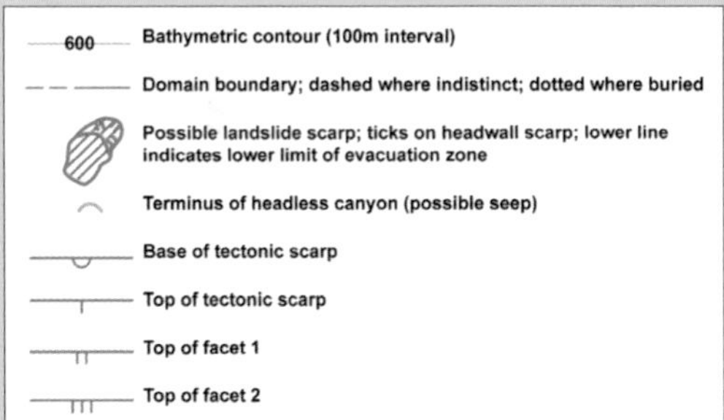

Geologic mapping of the seafloor in a GIS environment is being used to characterize the surface and subsurface geologic framework of the central California coastal region. Digital elevation models (DEMs) at the highest possible resolution were developed as a base for geomorphic interpretation of geologic features. These DEMs are based on reprocessing the individual surveys that compose the National Geophysical Data Center/National Oceanic and Atmospheric Administration (NGDC/NOAA) Coastal Relief Model. A slope model of the offshore study region was produced from a composite of this data and other available bathymetric DEMs using ArcGIS 3D Analyst. A geomorphic seafloor classification map was developed based on the slope model, published maps, available single-channel seismic reflection, and multibeam bathymetric data. Eight geomorphic domains were classified showing differing geologic, structural, and topographic characteristics along the continental slope and shelf.

Courtesy of Geomatrix Consultants, Pacific Gas and Electric.

Map of the Nicaraguan Volcanic Chain

Czech Geological Survey
Prague, Czech Republic
By Robert Tomas, Zuzana Krejci, Michaela Zemkova, Eva Zitova, Eva Kunceova, Petr Hradecky, and Jiri Sebesta

Contact
Dr. Robert Tomas
tomas@cgu.cz

Software
ArcGIS Desktop 9.1, Microstation 7, Adobe Illustrator CS

Printer
HP Designjet 5500

Data Source(s)
Geological data, natural hazards data

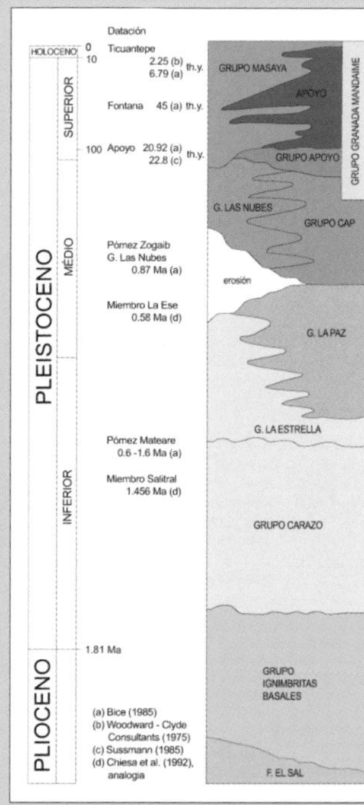

This map is based on detailed studies performed during 1997–2001 by the Czech Geological Survey and Instituto Nicaragüense de Estudios Territoriales (INETER) of Managua, Nicaragua, under the supervision of the Czech Ministry of Environment. It illustrates the main geological and geomorphological complexes and features in the Pacific zone of Nicaragua.

Until recently, the only existing regional photogeological map was based on data collected in 1972. The current data is based on newly performed field work and analytical results. Geological mapping was accompanied by the study of relief origin. The combination of both research methods was the basis for the compilation of the geological hazards assessment map.

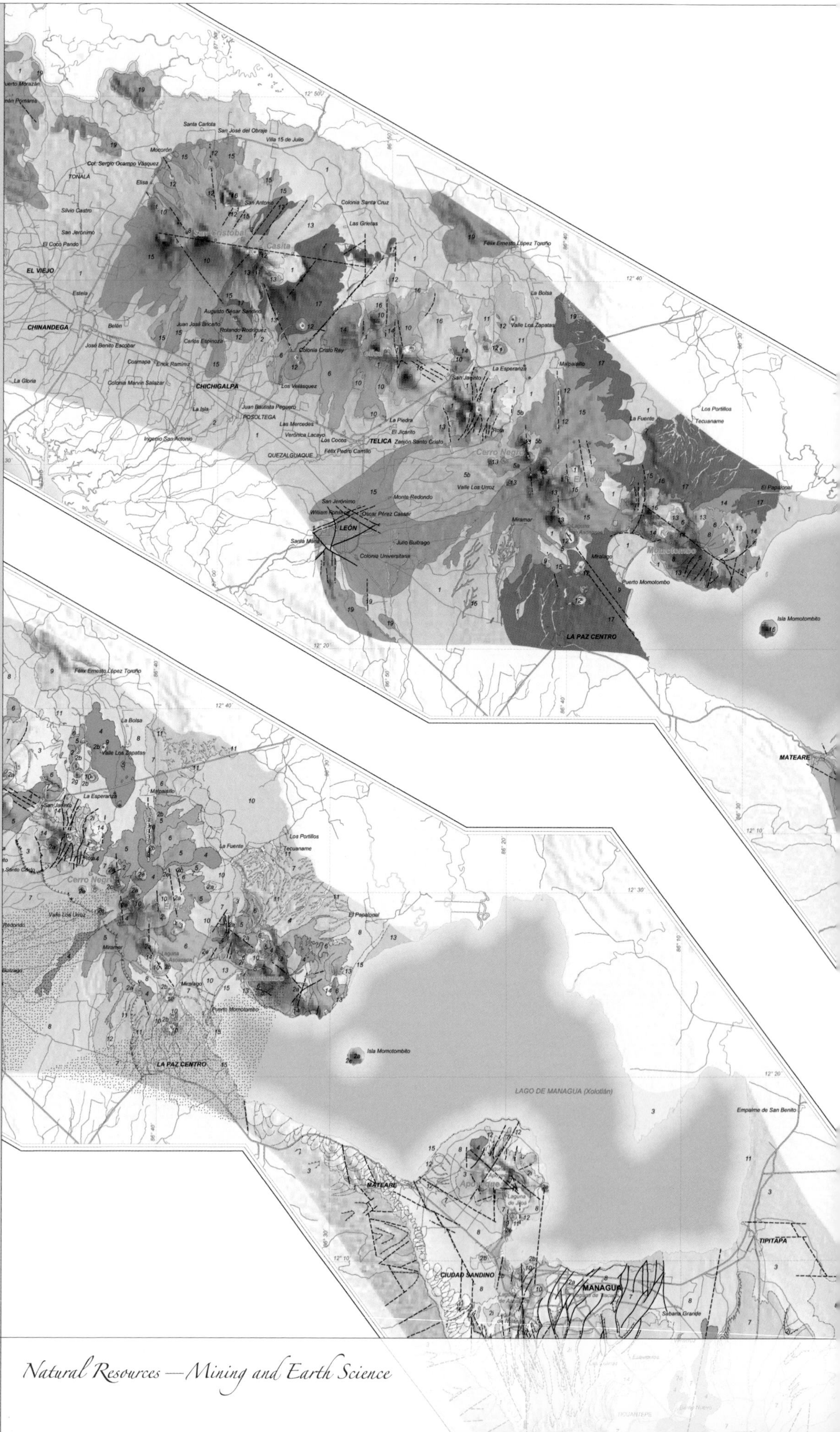

University of Texas Institute for Geophysics
Austin, Texas, USA
By Tim Whiteaker and Patricia Ganey-Curry

Contact
Timothy Whiteaker
twhit@mail.utexas.edu

Software
ArcGIS Desktop 9

Printer
HP Designjet 1055cm

Data Source(s)
Published well-log cross-section sets, Texas Bureau of Economic Geology, Louisiana Geological Survey, Minerals Management Service, UTIG seismic lines, MMS and Paleo-Data paleontologic reports, University of Texas theses and dissertations.

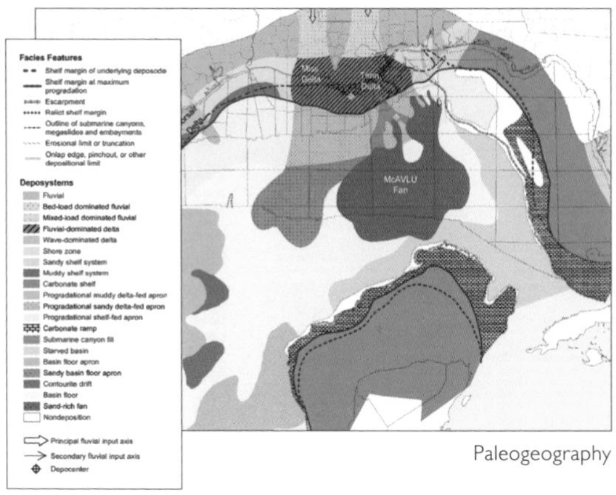

Paleogeography

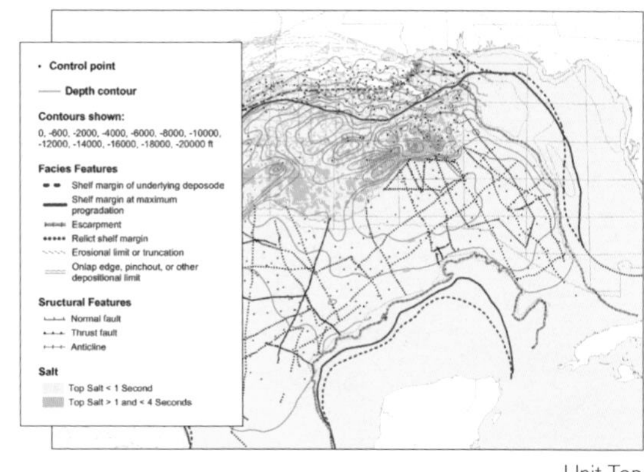

Unit Top

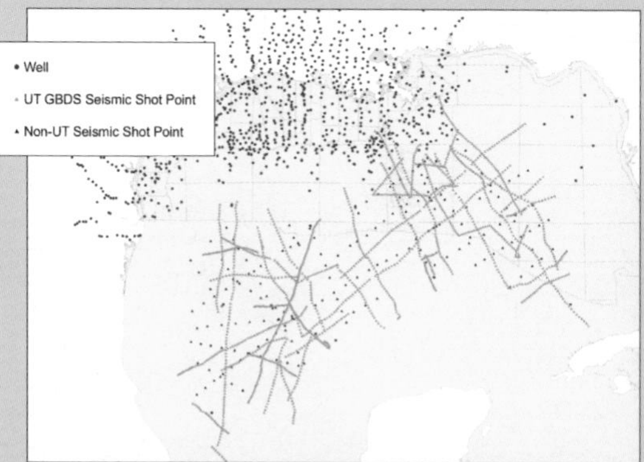

Gulf Basin Depositional Synthesis Points

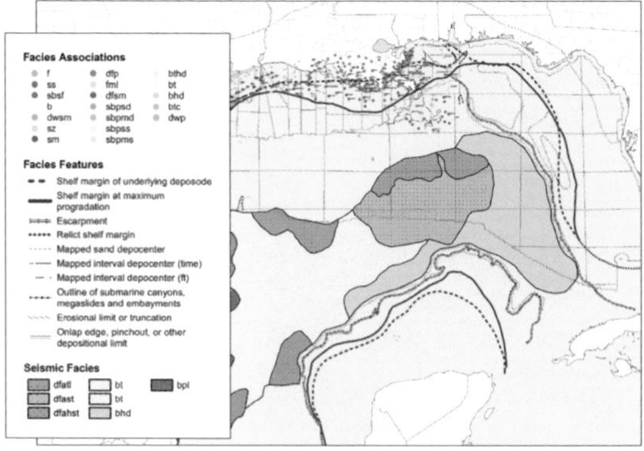

Depositional Framework

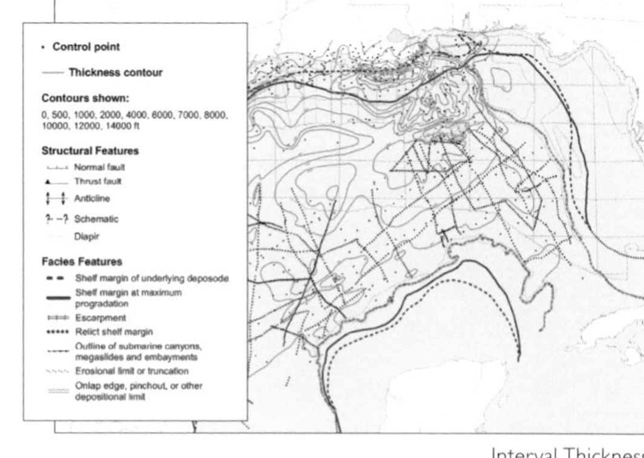

Interval Thickness

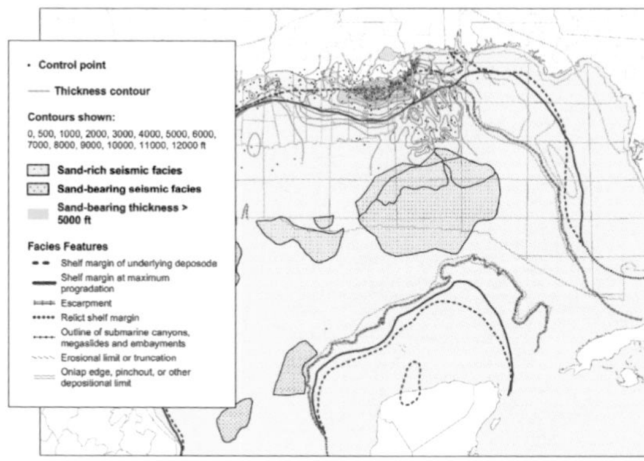

Sand Bearing Interval Thickness

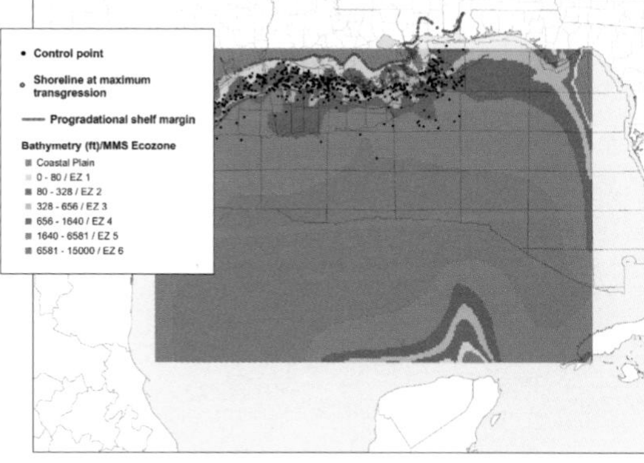

2D Paleobathymetry

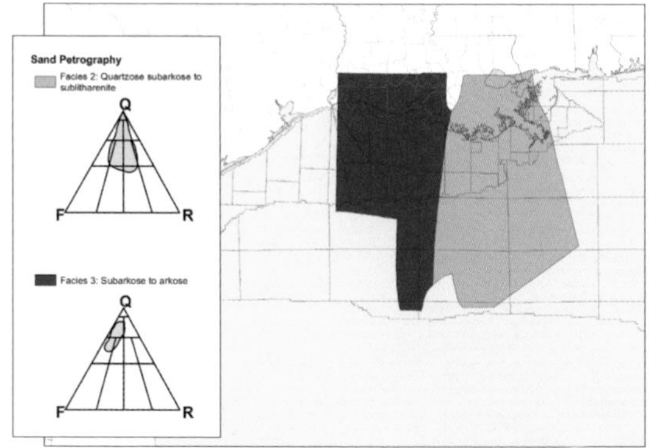

Petrographic Facies

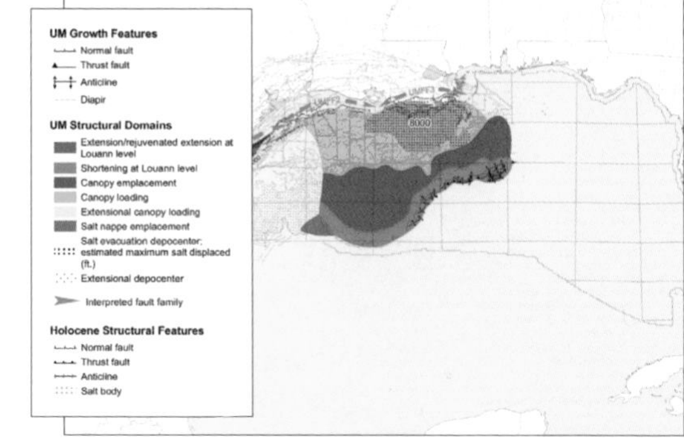

Structural Features

The Gulf Basin Depositional Synthesis (GBDS) Project is a joint industry-sponsored research project to provide a complete and comprehensive picture of basin-scale sedimentation as a tool for exploration efforts in the Gulf of Mexico. The project contributes new information to the understanding of the geologic history of the Gulf of Mexico basin. The database has integrated well data from the onshore, shelf, and upper slope areas with seismic data from the deep Gulf of Mexico basin. Data interpretations were also made by scientists at the University of Texas Institute for Geophysics (UTIG). The database is implemented as a personal geodatabase and makes use of subtypes to distinguish data across each stratigraphic sequence.

Courtesy of Tim Whiteaker and Patricia Ganey-Curry, University of Texas at Austin.

A Cartographic Library of Regional Oil and Gas Interests on the North Slope of Alaska

State of Alaska, Division of Oil and Gas
Anchorage, Alaska, USA
By Christina Beaty

Contact

Christina Beaty
christina_beaty@dnr.state.ak.us

Software

ArcGIS Desktop, Adobe Illustrator, other GIS software

Printer

HP Designjet 5500

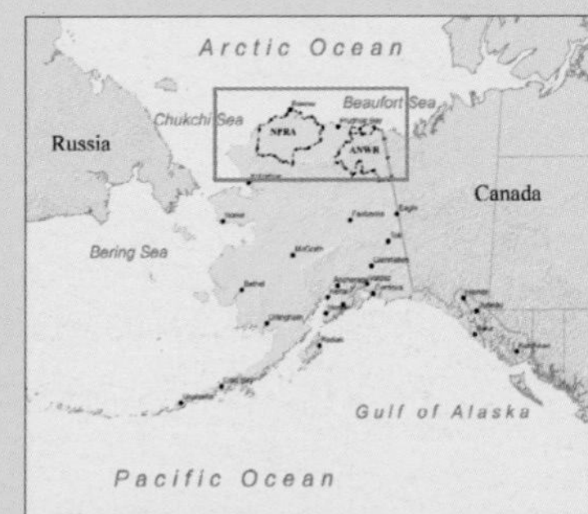

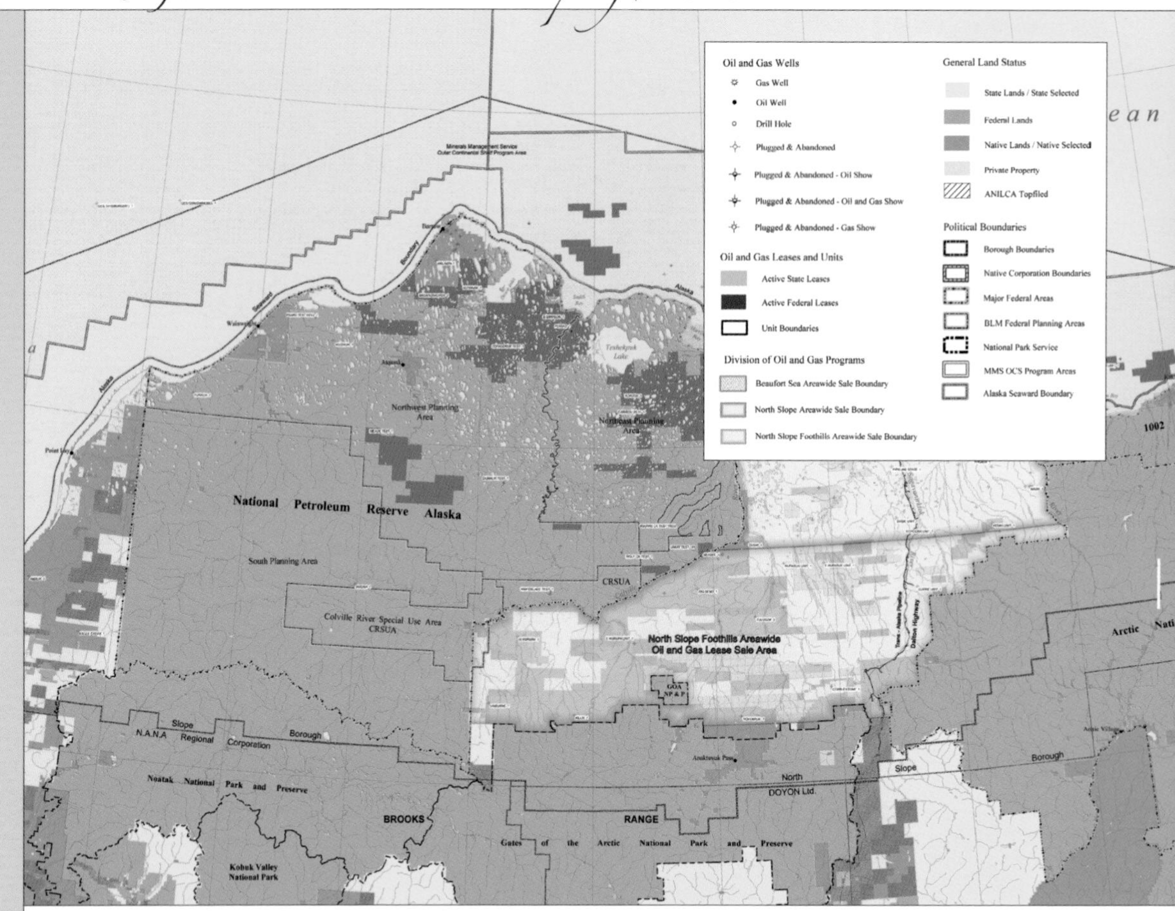

Plate 1: Oil and Gas Programs for the North Slope of Alaska

Plate 2: Regional Geology of the North Slope of Alaska

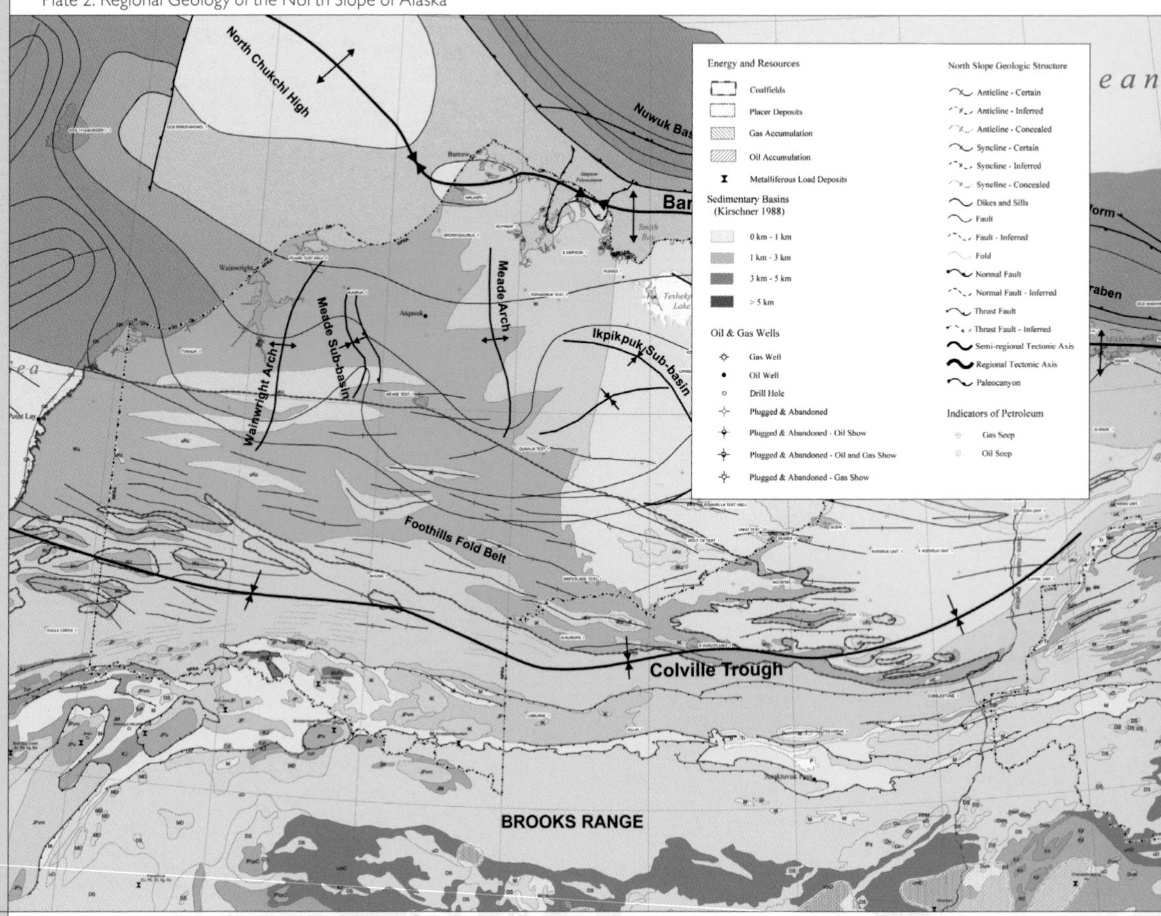

To many potential explorers, the North Slope of Alaska is an intimidating wasteland, far from infrastructure and support, and acquiring data about this region can be frustrating. By compiling and providing general data, regional geology, political boundaries, and state programs, the Division of Oil and Gas is extending valuable research tools. This map series consists of four 80×36-inch maps to be used as posters for quick visual reference and to inspire discussion about North Slope resources. Included are some well-known areas such as the Arctic National Wildlife Refuge, National Petroleum Reserve–Alaska, the Brooks Range, Beaufort Sea, Chukchi Sea, and Prudhoe Bay.

Map Plate 1 focuses on oil and gas sale areas, leased land, oil and gas units, land ownership, borough boundaries, and federal boundaries. Regional geology is the focus of Map Plate 2. It shows surficial rock classification, large-scale structure, energy resource accumulations, and basin information. Geologic cross sections across main areas of interest and a geologic discussion enhance the map. Map Plate 3 displays publicly available and brokered seismic data locations. Three regional lines were selected and shown on the poster to illustrate the data quality of certain surveys and trends that are visible in the seismic data. Map Plate 4 shows four main data frames: publicly available gravity data, publicly available magnetic data, a regional gravitational anomaly, and a regional magnetic anomaly.

Courtesy of the State of Alaska, Division of Oil and Gas.

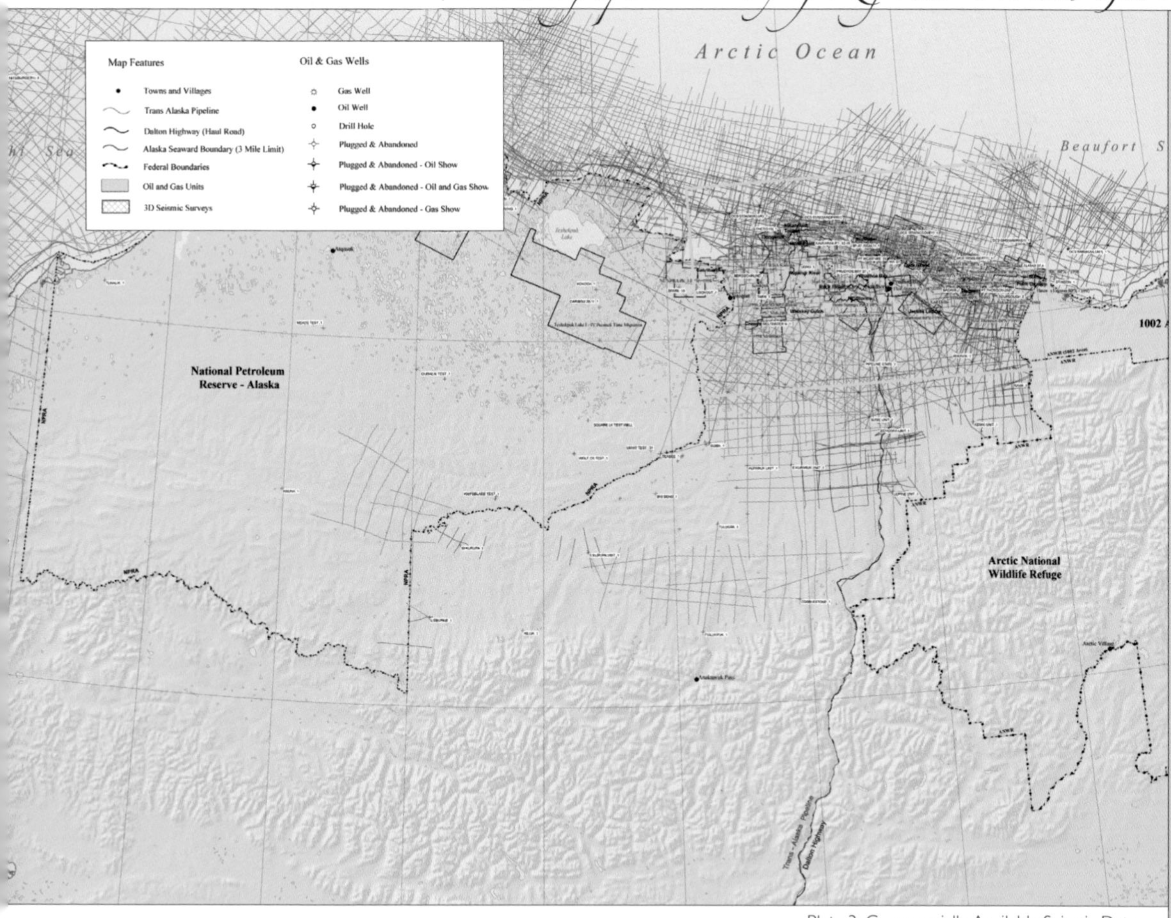

Plate 3: Commercially Available Seismic Data

Plate 4: Publicly Available Gravity Data

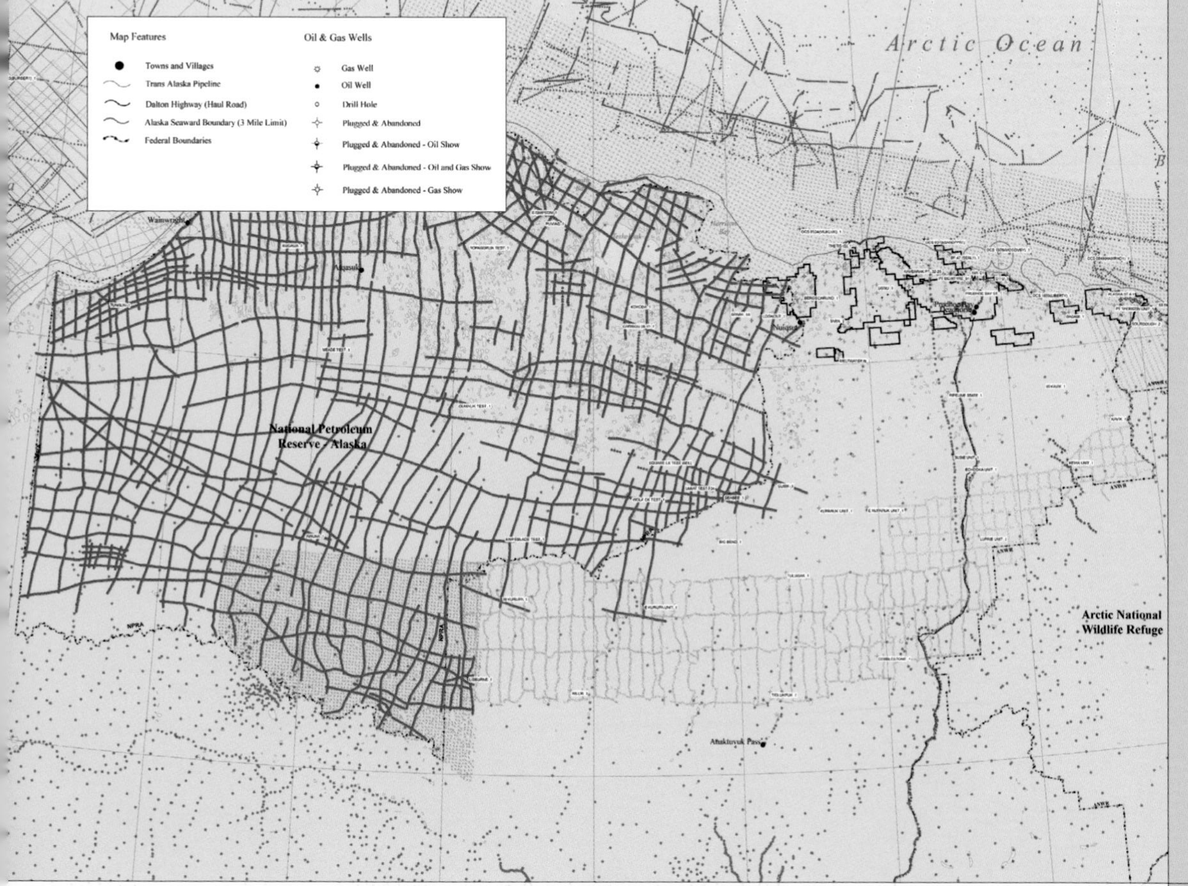

Gravity Data Available from Carson Services			
Symbol	**Survey Name**	**Location**	**# of Data Points**
——	Carson Survey 1	ANWR	unknown
——	Carson Survey 2	ANWR	unknown
Gravity Data Available from EDCON			
Symbol	**Survey Name**	**Location**	**# of Data Points**
——	Chukchi Sea Barrow Arch 1985	Chukchi Sea/Arctic Ocean	unknown
——	Chukchi Sea 1982-1985	Chukchi Sea/Arctic Ocean	unknown
——	Chukchi Sea 1984	Chukchi Sea/Arctic Ocean	unknown
——	North Chukchi Sea Ice	Arctic Ocean	517
+	Beaufort Sea Coastal Survey	Beaufort Sea Coast	123
x	Beaufort Sea Ice	Beaufort Sea/Arctic Ocean	375
x	Southern NPRA	Southern NPRA	3626
Gravity Data Available from PhotoGravity			
Symbol	**Survey Name**	**Location**	**# of Data Points**
x	Photo Gravity	Several Locations	17,583
Gravity Data Available from the Department of Defense			
Symbol	**Survey Name**	**Location**	**# of Data Points**
•	Compilation of Surveys	North Slope	74,436
Gravity Data Available from USGS			
Symbol	**Survey Name**	**Location**	**# of Data Points**
•	Compilation of Surveys	North Slope	83,703
Note			
●	Some data is available from both the DOD and the USGS. This is especially true in NPRA. Although these data are spatially identical, processing parameters could vary.		

Wood Mackenzie, Ltd.
Edinburgh, UK
By Stephen Bull

Contact
Stephen Bull
stephen.bull@woodmac.com

Software
ArcGIS Desktop

Printer
HP Designjet 750c

Data Source(s)
Wood Mackenzie, Ltd.

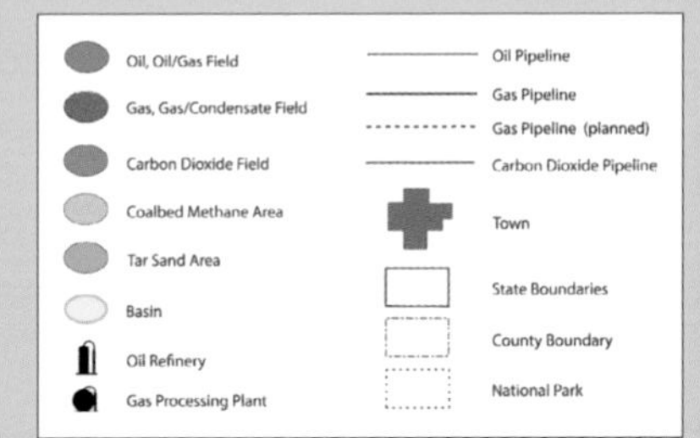

- Oil, Oil/Gas Field
- Gas, Gas/Condensate Field
- Carbon Dioxide Field
- Coalbed Methane Area
- Tar Sand Area
- Basin
- Oil Refinery
- Gas Processing Plant
- Oil Pipeline
- Gas Pipeline
- Gas Pipeline (planned)
- Carbon Dioxide Pipeline
- Town
- State Boundaries
- County Boundary
- National Park

Wood Mackenzie has produced a detailed wall map of the U.S. Rocky Mountains area showing the latest information about the oil and gas infrastructure in the region. The map shows active oil and gas fields and basins, refineries, coal-bed methane areas, and carbon dioxide fields. It also gives locations of existing and planned pipelines.

In addition, an extensive legend charts top companies by gross lease holdings by basin; remaining reserves by basin and hydrocarbon type; estimated production by basin; and reserves, value, and production by company.

The map allows clients to get a rapid appreciation of the scale and nature of the oil and gas business in the Rockies area, the location of the main oil and gas projects, the level of infrastructure, and the trends in exploration and production.

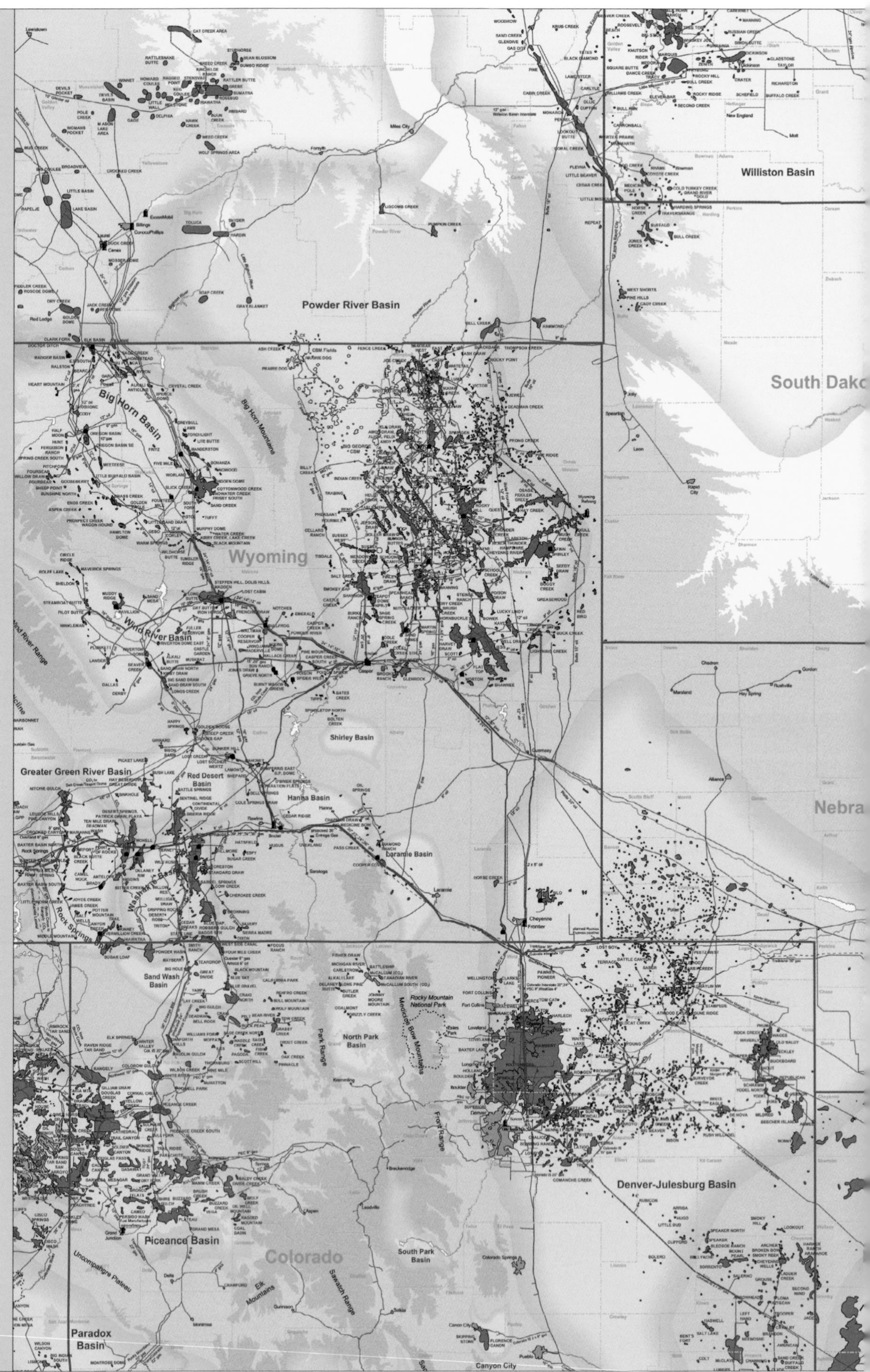

Mammoth Community Water District

Mammoth Lakes, California, USA
By Justin Mulbay and Tom McCarthy

Contact
Justin Mulbay
jmulbay@mcwd.dst.ca.us

Software
ArcEditor, ArcGIS 3D Analyst, Surfer, Adobe Photoshop

Printer
HP Designjet 5000ps

Data Source(s)
Mammoth Community Water District,
National Agriculture Imagery Program (NAIP), USGS

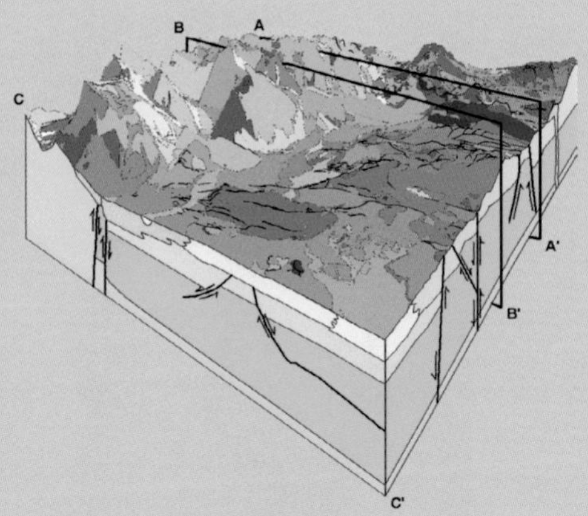

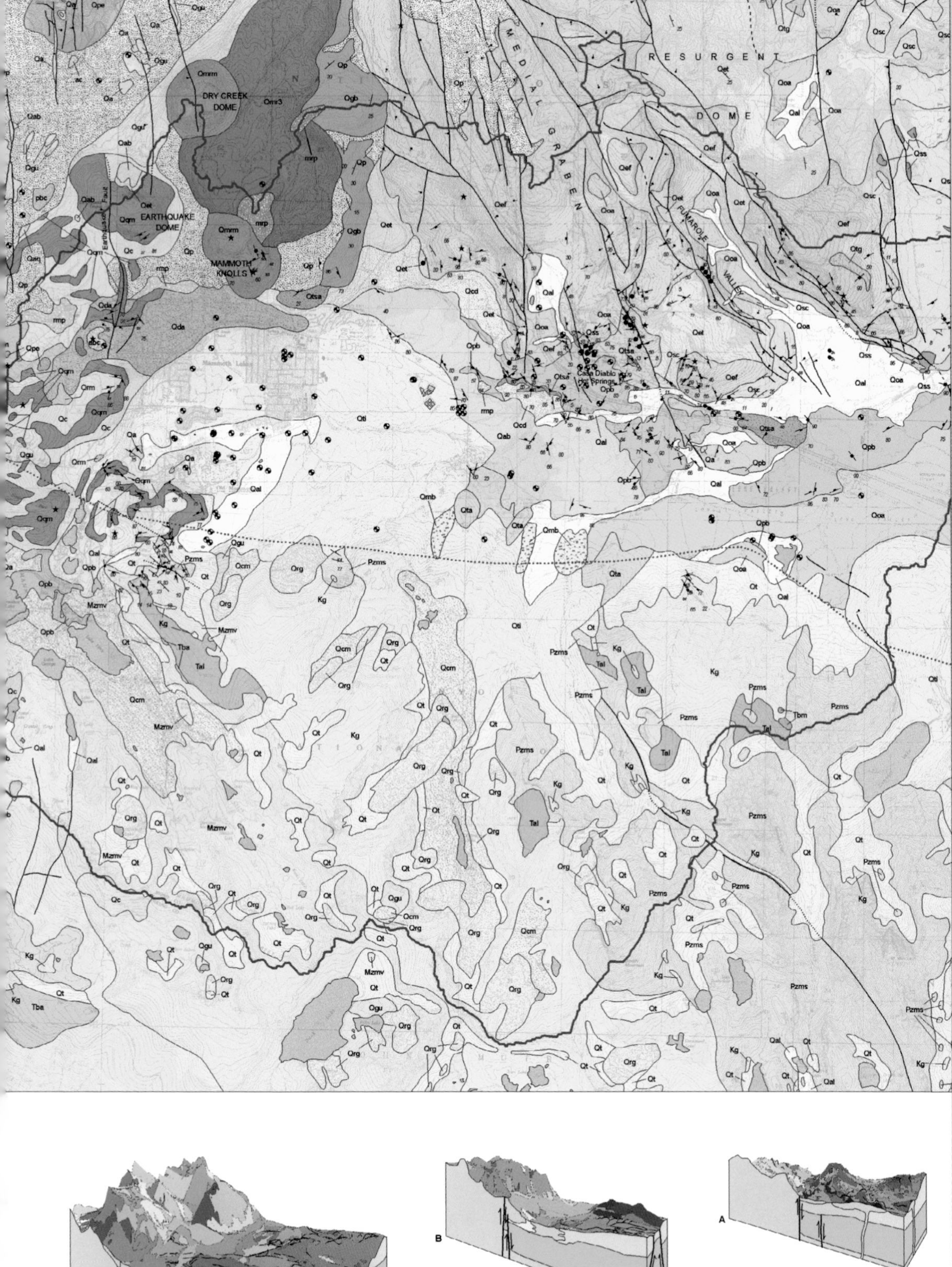

To date, there has been no scientific determination of the safe yield of the Mammoth Basin in Mono County, California. The potential annual volume of groundwater extraction is very uncertain, yet it plays a vital role in the Mammoth Community Water District (MCWD) water supply, thereby making high-certainty water resources planning difficult. The MCWD has initiated a rigorous groundwater monitoring plan to observe groundwater dynamics within the basin and gather data to create a numerical model. Specific goals include defining how Mammoth Basin hydraulically operates; the interconnection of the upper and lower aquifers; the level of connection between Mammoth Creek and the basin groundwater system; and the volume that can regularly be pumped out of the basin without undesired effects.

The ultimate objective of this modeling project is to reduce uncertainty in the supply of groundwater resources by developing a management tool to assist MCWD in predicting available groundwater resources while managing storage for future needs and protection of the aquifers.

Courtesy of Mammoth Community Water District.

LEGEND
Bedrock
Alluvium Low
Alluvium High
Till
Rhyolite
Volcanic Low
Volcanic High

Errol L. Montgomery & Associates
Tucson, Arizona, USA
By Carolyn Lambert and Hale Barter

Contact

Carolyn Lambert
clambert@elmontgomery.com

Software
ArcGIS Desktop 9.1, ArcGIS Spatial Analyst

Printer
HP Designjet 800

Data Source(s)
ADWR, USGS

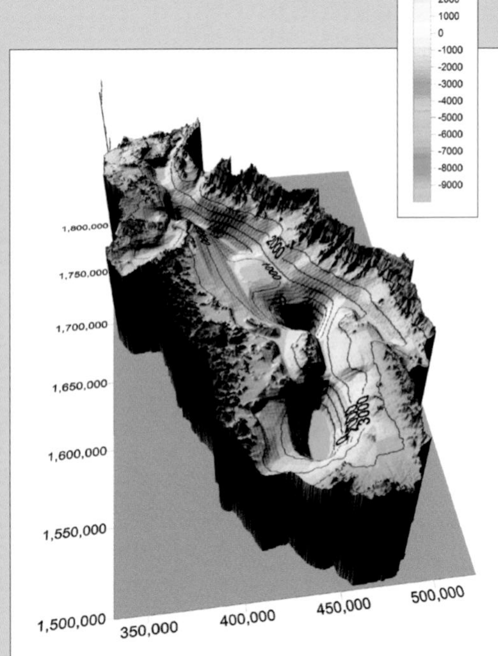

Altitude of Basin Bedrock

As part of a basin-wide hydrogeologic investigation, this groundwater flow model was developed for an area in northwestern Arizona. The model is used as a tool to predict the impact of potential groundwater demands in the basin. In the past, hydrogeologic data used to construct the model would be prepared in spreadsheets and ASCII data files before importing the properties into the pre- and post-processing software called Groundwater Vistas. Groundwater Vistas now has the ability to use shapefiles for all modeling datasets. ArcGIS is used to provide a central location for spatial data storage, preparation, surface calculations, and analysis of the model results.

Courtesy of Errol L. Montgomery & Associates.

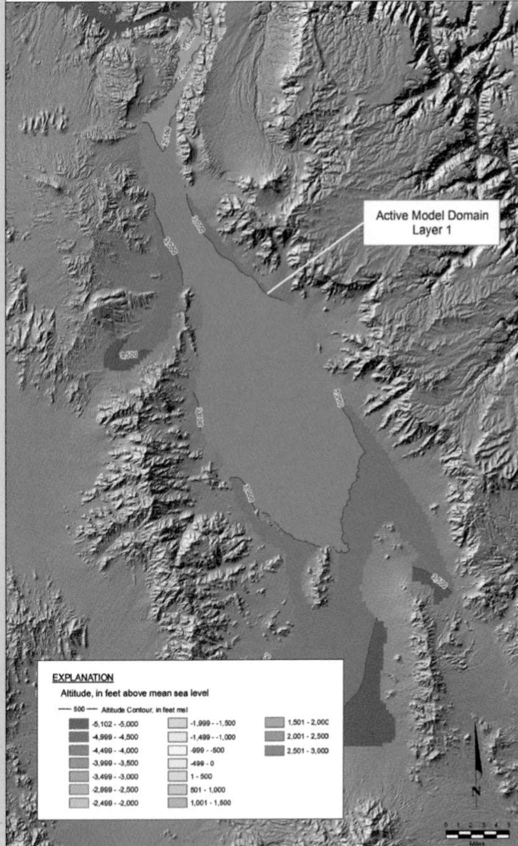

Land Surface Altitude

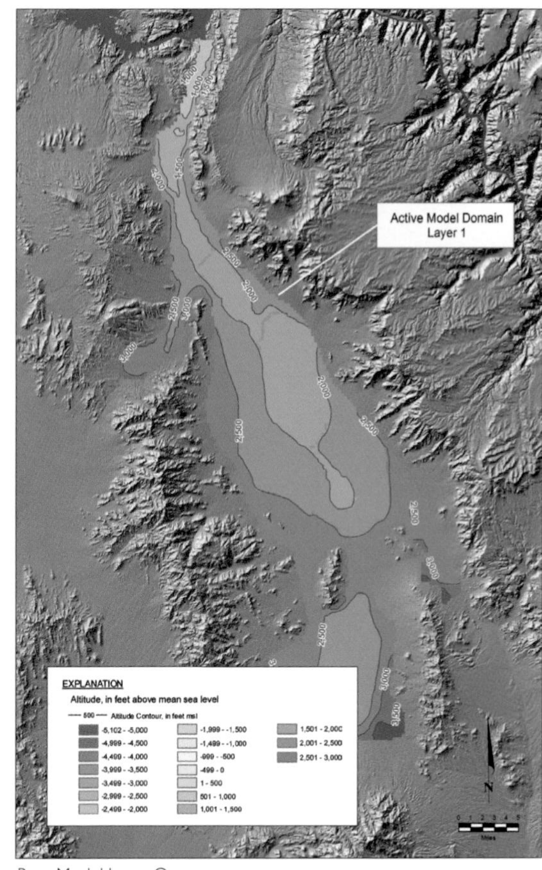

Base Model Layer One

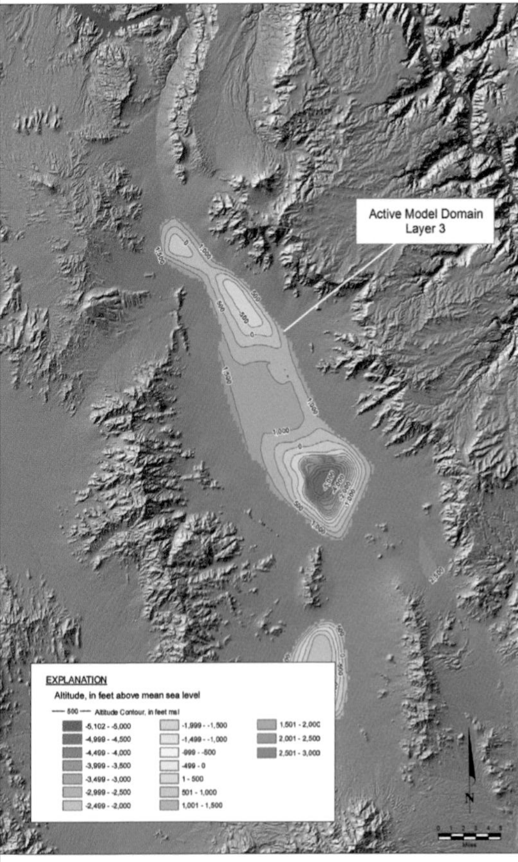

Base Model Layer Three

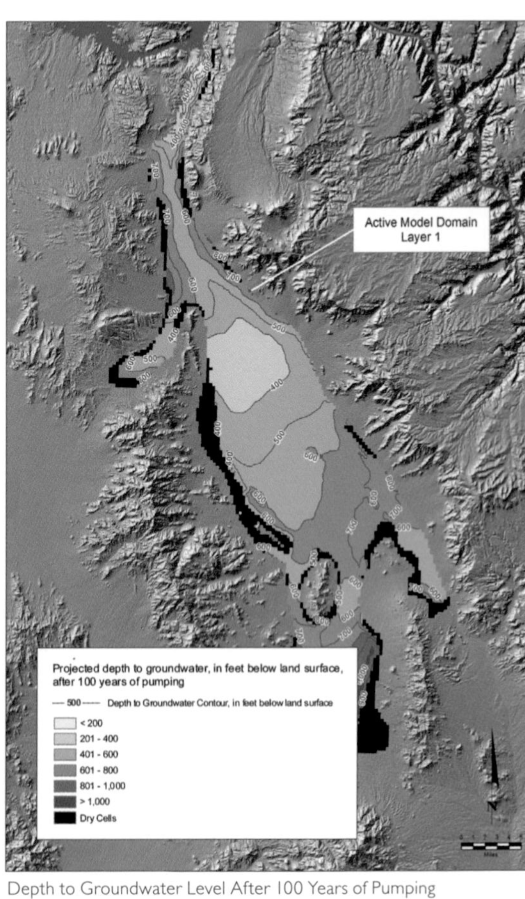

Depth to Groundwater Level After 100 Years of Pumping

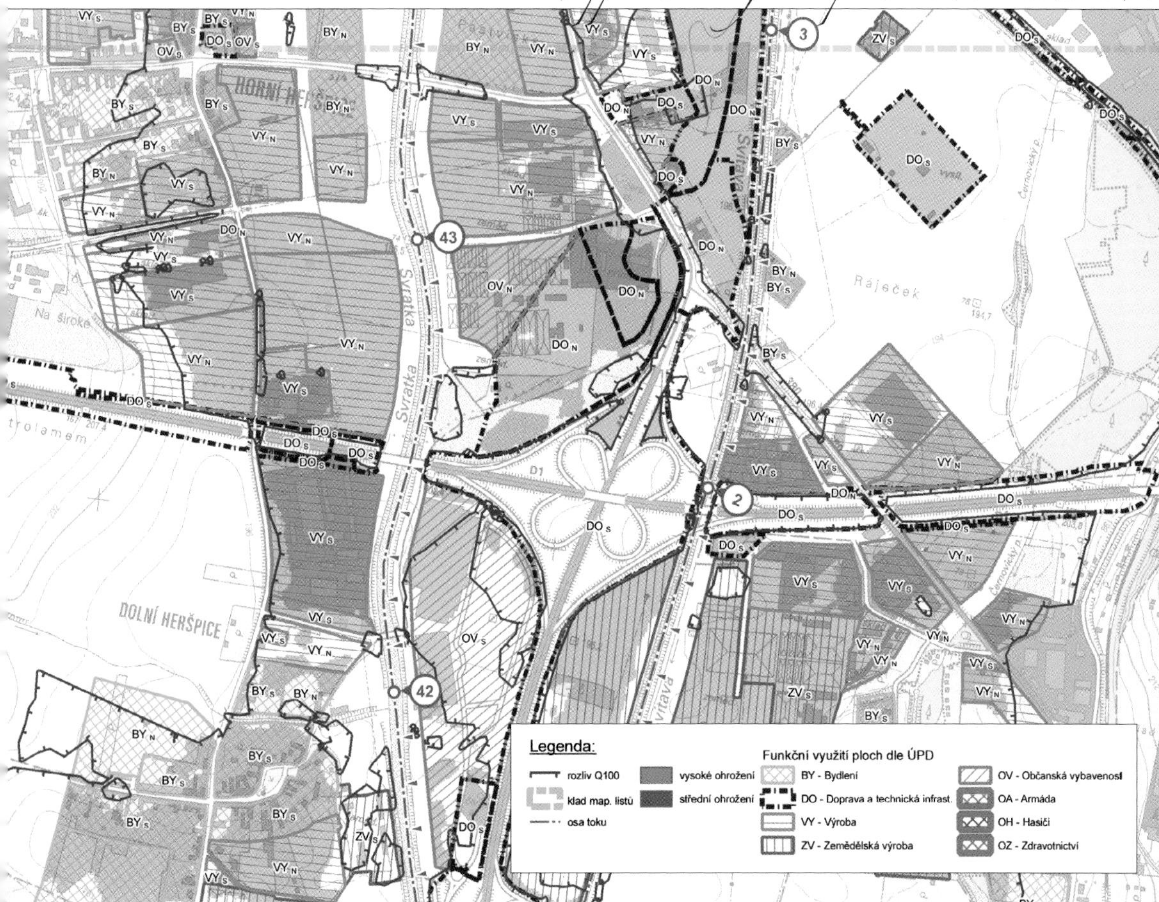

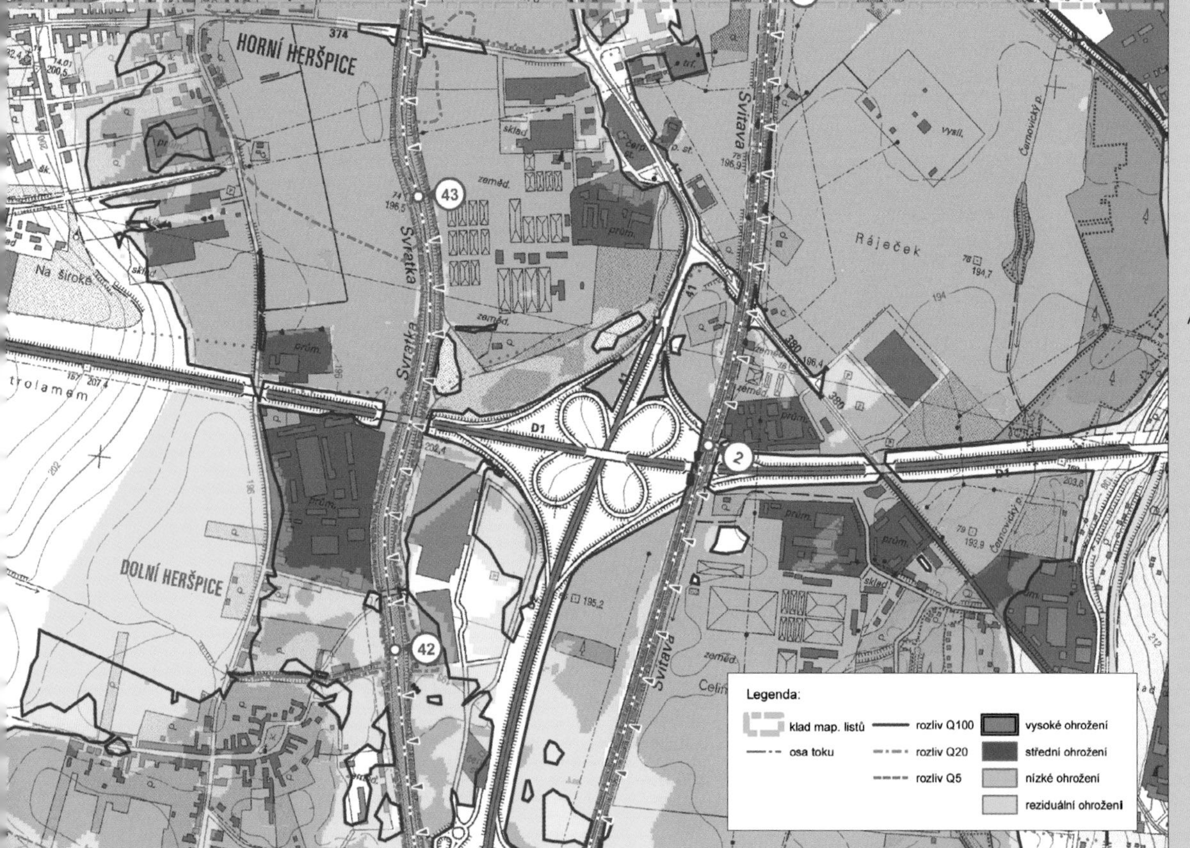

Brno University of Technology
Brno, Czech Republic
By Aleš Dráb, Jiirí Cejp, Tomáš Julínek, and Helena Koutková

Contact
Aleš Dráb
drab.a@fce.vutbr.cz

Software
ArcGIS Desktop 9.1

Printer
HP Designjet 800

Data Source(s)
Ríha, J. et al. 2005, Prague Flood Protection Measures—Groundwater Modelling Study, Brno University of Technology, FCE, Department of Water Structures

These maps illustrate the application of ArcGIS deterministic and statistical interpolation methods in practical problems of groundwater flow. The methods are used in specific procedures of data preprocessing, as well as postprocessing of data calculated by numerical models of groundwater flow. The basic principle in applying interpolation methods rests in creating a continuous distribution of examined variables (e.g., groundwater surface, impermeable layer) based on values of discrete data (e.g., boreholes, nodes of calculation grid) measured in the area of interest. The project addresses mainly interpolation methods included in the ArcGIS 3D Analyst, ArcGIS Spatial Analyst, and ArcGIS Geostatistical Analyst extensions. The project acknowledges the study "Analysis of Risks Associated with Groundwater Regime Changes in Extreme Hydrological Situations" (Grant Agency of Czech Republic No. 103/06/0595).

Courtesy of Aleš Dráb, Jiirí Cejp, Tomáš Julínek, and Helena Koutková, Brno University of Technology.

Using ArcGIS to Monitor the Effect of Water Table Fluctuations on the Behavior of Non-Aqueous Phase Petroleum

Langan Engineering and Environmental Services, Inc.
Doylestown, Pennsylvania, USA
By Brett R. Milburn

Contact
Brett Milburn
bmilburn@langan.com

Software
ArcGIS Desktop 9.1

Printer
HP Designjet 1055cm

Data Source(s)
GPS, surveying, aerial photography

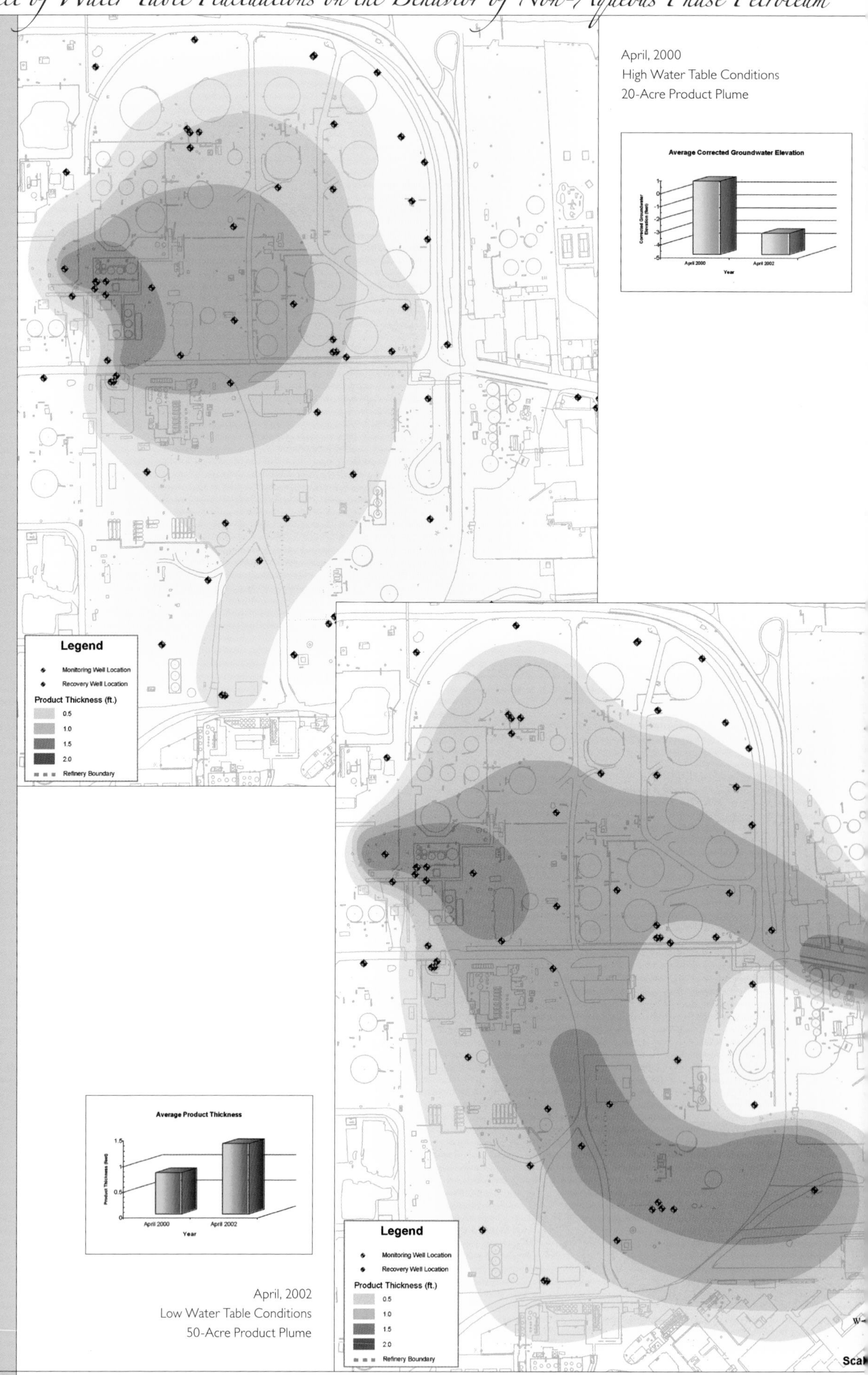

April, 2000
High Water Table Conditions
20-Acre Product Plume

Average Corrected Groundwater Elevation

Corrected Groundwater Elevation (feet)

April 2000 April 2002

Year

Legend

Monitoring Well Location
Recovery Well Location

Product Thickness (ft.)
0.5
1.0
1.5
2.0
Refinery Boundary

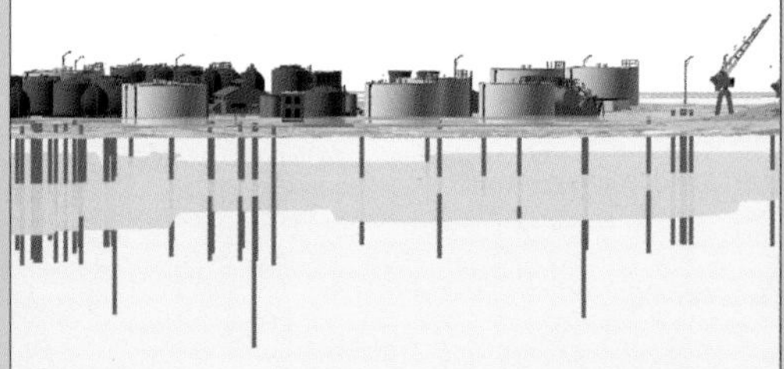

Portion of 3D Model Depicting the April 2002 Product Plume

A conceptual model was created using ArcGIS to analyze and illustrate the behavior of light non-aqueous-phase petroleum (LNAPL) in a heterogeneous aquifer system under an active petroleum refinery. The information obtained was used extensively to demonstrate the influence of water table fluctuations on apparent product thickness in site monitoring and recovery wells. The ArcView Spatial Analyst extension was used to contour groundwater elevations and LNAPL thickness as recorded during site-gauging events. The charts and contour maps created were used to compare groundwater elevation and product thickness data for the same month exactly two years apart (April 2000 and April 2002). Data collected from April 2000 was recorded during a period of long-term high water table conditions. By contrast, data from April 2002 was collected during long-term low water table conditions, a result of a severe regional drought. To further visualize the overall extent and thickness of the product plume across the site during periods of high- and low-water conditions, a three-dimensional model using the ArcView 3D Analyst extension was created in ArcGIS. Currently, the model is being used to visualize product behavior, analyze seasonal and climatic effects to anticipate potential changes in product recovery rates, and support the design of appropriate remedial systems across the site.

Courtesy of Langan Engineering and Environmental Services, Inc.

Average Product Thickness

Product Thickness (feet)

1.5
1
0.5
0

April 2000 April 2002

Year

April, 2002
Low Water Table Conditions
50-Acre Product Plume

Legend

Monitoring Well Location
Recovery Well Location

Product Thickness (ft.)
0.5
1.0
1.5
2.0
Refinery Boundary

Santa Ana Watershed Project Authority
Riverside, California, USA
By Mario Chavez, Pete Vitt, and Jerry Oldenburg

Contact
Peter Vitt
pvitt@sawpa.org

Software
ArcGIS Desktop

Printer
HP Designjet 4000

Data Source(s)
USGS, SAWPA

The Santa Ana River Trail Partnership consists of cities and counties within the Santa Ana Watershed in Southern California, as well as governmental agencies and environmental groups dedicated to completion of the Santa Ana River Trail from the mountains to the ocean. This map depicts their plan for completion of the trail, as well as other regional trails and camping and recreational sites along the Santa Ana River Corridor. Identified on the map are important regional trail linkages and recreational opportunities in a rapidly urbanizing region.

Courtesy of Santa Ana Watershed Project Authority.

City of Salinas Zoning Map

City of Salinas

Salinas, California, USA
By Charles Lerable

Contact

Charles Lerable
chuck@ci.salinas.ca.us

Software

ArcGIS Desktop, ArcGIS 3D Analyst

Printer

HP DesignJet 5000

Data Source(s)

City of Salinas

Downtown Neighborhood Area
Downtown Core Overlay Area
E MARKET ST
MARKET ST
W ALISAL ST
JOHN ST
FRONT ST
MAIN ST
CITY HALL
MONTEREY COUNTY OFFICES & COURTS
SALINAS HIGH SCHOOL

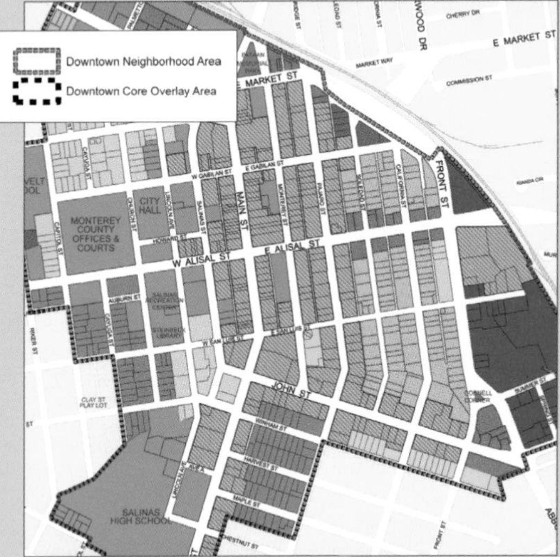

Central City Overlay

Zoning is arguably the most important tool for regulating land development. It is also a key regulatory tool for implementing a city's comprehensive general plan. In Salinas, California, as in most municipalities and counties, zoning involves establishing regulations that control the type of development (e.g., residential, commercial, or industrial) and property development standards (e.g., building placement, parking, signage).

Salinas is divided into 25 zones, or zoning districts, each with different and sometimes overlapping regulations. Communicating this information to the public is very important because zoning affects business expansion, establishes neighborhood standards, and otherwise provides a predictable basis for community change and the regulation of public nuisance.

This zoning map consists of zoning districts and "overlay districts" where additional special regulations apply pertaining to flood protection, airport overflights, and architectural design review. These zones are rendered using a combination of colors, polygon boundaries, and map insets, all with the purpose of providing a single and uncomplicated map of where zoning regulations apply.

Courtesy of the City of Salinas, California.

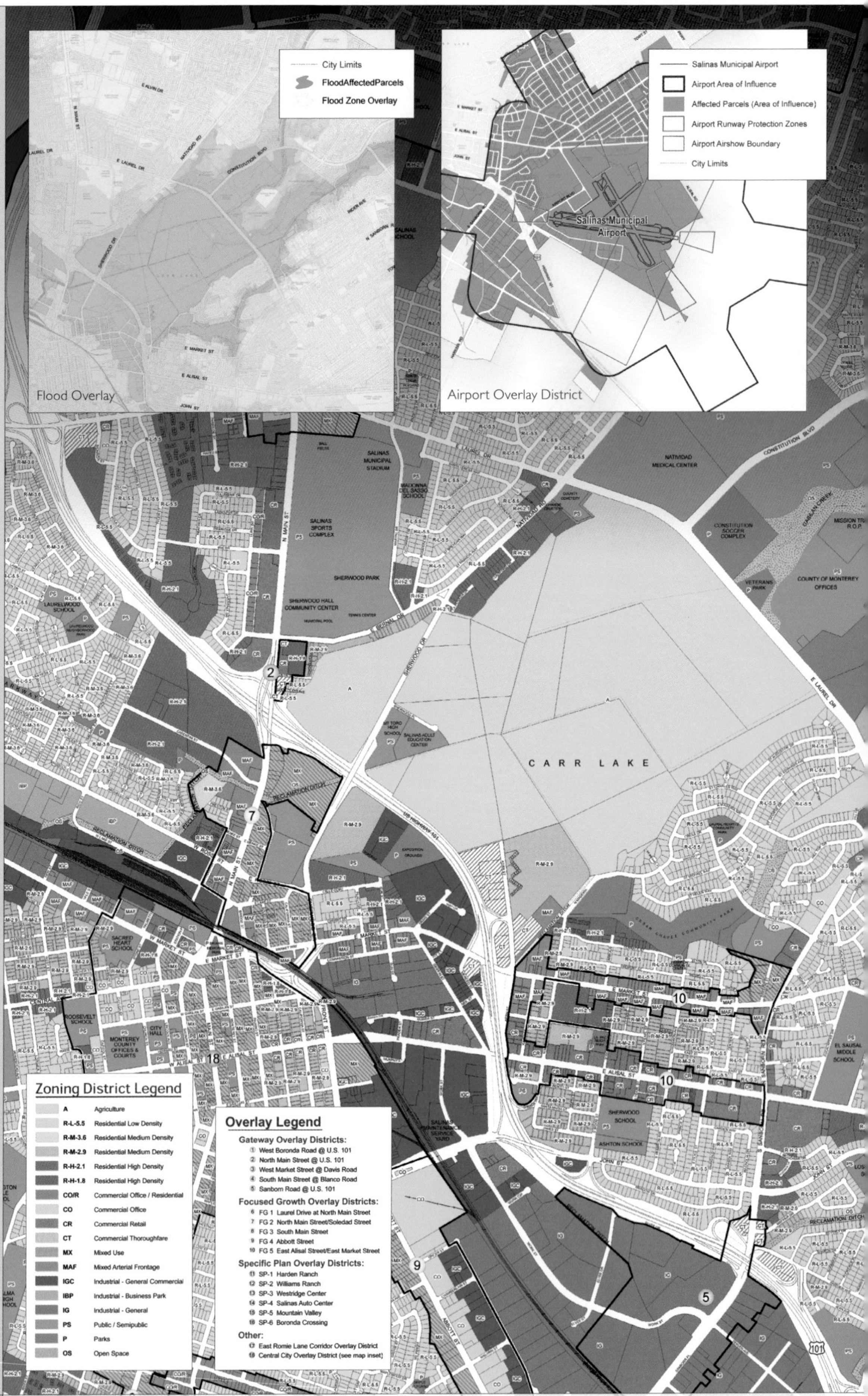

City Limits
FloodAffectedParcels
Flood Zone Overlay
Flood Overlay
Salinas Municipal Airport
Airport Area of Influence
Affected Parcels (Area of Influence)
Airport Runway Protection Zones
Airport Airshow Boundary
City Limits
Airport Overlay District
CARR LAKE
Zoning District Legend
A Agriculture
R-L-5.5 Residential Low Density
R-M-3.6 Residential Medium Density
R-M-2.9 Residential Medium Density
R-H-2.1 Residential High Density
R-H-1.8 Residential High Density
CO/R Commercial Office / Residential
CO Commercial Office
CR Commercial Retail
CT Commercial Thoroughfare
MX Mixed Use
MAF Mixed Arterial Frontage
IGC Industrial - General Commercial
IBP Industrial - Business Park
IG Industrial - General
PS Public / Semipublic
P Parks
OS Open Space
Overlay Legend
Gateway Overlay Districts:
1 West Boronda Road @ U.S. 101
2 North Main Street @ U.S. 101
3 West Market Street @ Davis Road
4 South Main Street @ Blanco Road
5 Sanborn Road @ U.S. 101
Focused Growth Overlay Districts:
6 FG 1 Laurel Drive at North Main Street
7 FG 2 North Main Street/Soledad Street
8 FG 3 South Main Street
9 FG 4 Abbott Street
10 FG 5 East Alisal Street/East Market Street
Specific Plan Overlay Districts:
11 SP-1 Harden Ranch
12 SP-2 Williams Ranch
13 SP-3 Westridge Center
14 SP-4 Salinas Auto Center
15 SP-5 Mountain Valley
16 SP-6 Boronda Crossing
Other:
17 East Romie Lane Corridor Overlay District
18 Central City Overlay District (see map inset)

Modeling Affordable Home Ownership in South Australia

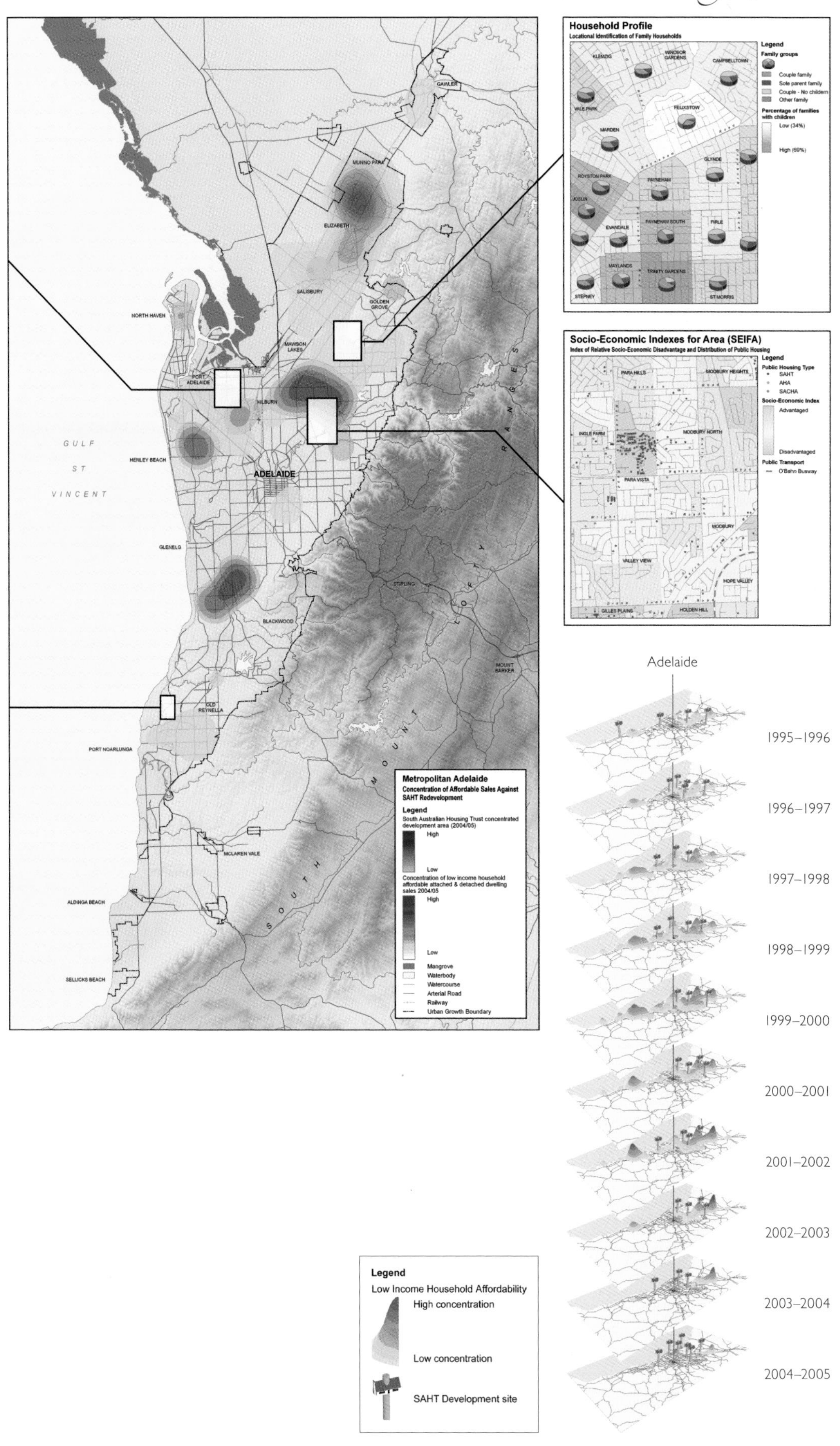

Department for Families and Communities
Adelaide, Australia
By Gary Maguire and Jarrod Gitsham

Contact
Gary Maguire
gary.maguire@dfc.sa.gov.au

Software
ArcGIS Desktop 9.1, ArcGIS Spatial Analyst, ArcGIS 3D Analyst, Adobe Photoshop CS2

Printer
HP Designjet 5000ps

Data Source(s)
DFC, DEH, DTEI, ABS Census

Home ownership offers important benefits that enable a household to build wealth through equity appreciation, and it is generally considered a secure form of housing. South Australia has seen significant increases in house sale prices over the past decade, leaving home ownership out of reach for many low-income households. This is particularly evident within the metropolitan Adelaide area where affordable home ownership opportunities have dramatically diminished and moved toward urban fringe areas, which typically are not well serviced and offer poor housing quality. GIS has provided the tools to visualize the affordable-dwelling sale trends and the impacts that the South Australian Housing Trust redevelopments are having on low-income households and their communities.

Courtesy of Department for Families and Communities, South Australia.

Center for Spatial Information Science, University of Tokyo
Kashiwa, Chiba Prefecture, Japan
By Akiko Takahashi and Hideto Satoh

Contact
Akiko Takahashi, Hideto Satoh
akuri@csis.u-tokyo.ac.jp, hideto@csis.u-tokyo.ac.jp

Software
ArcGIS Desktop 9, ArcGIS 3D Analyst

Printer
Canon W7200

Data Source(s)
Sinfonica Japan; Ministry of Land Infrastructure and Transport; Geographical Survey Institute; Zenrin Co., Ltd.

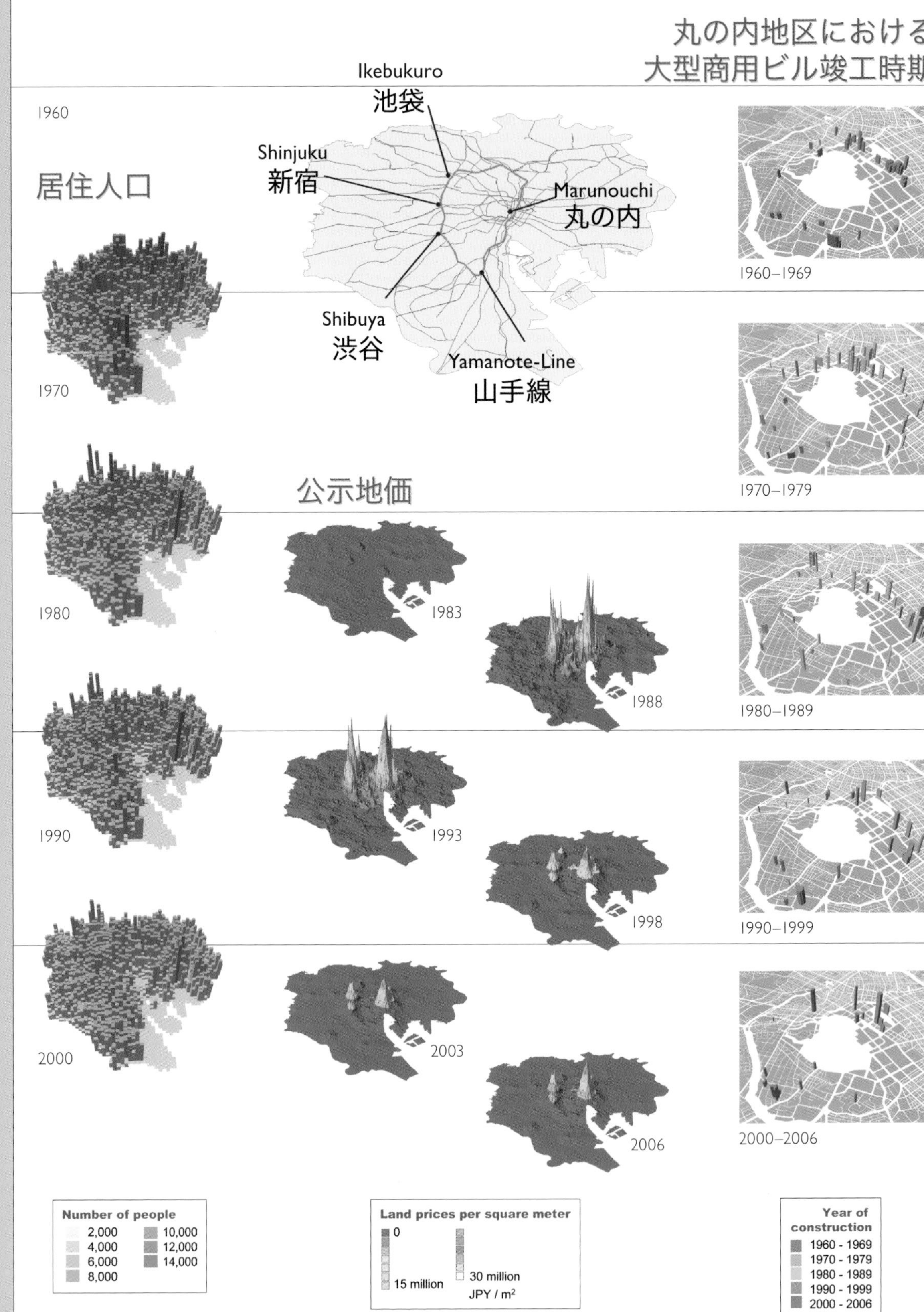

These maps show the changes in population, land prices, and number of large office buildings constructed in the Tokyo metropolitan area from the 1960s to today. From the late 1980s to the early 1990s, Japan experienced a bubble economy where the price of real estate escalated dramatically. Tokyo has a center named Marunouchi and three subcenters—Shinjuku, Ikebukuro, and Shibuya. Land prices in these areas were extremely high during the bubble economy, visible in the land price "mountain" in the middle column.

As a result, the distribution of office building construction changed. Large office buildings were built in central Marunouchi in the 1960s and 1970s. From the 1980s to the 1990s, building construction moved to areas around the Marunouchi district. After the bubble burst, land prices came down and the construction of office buildings returned to central Marunouchi. The residential population also changed over this period. The national population peaked in 2005, but in the Tokyo metropolitan area, it has been decreasing since the 1970s.

Courtesy of Center for Spatial Information Science, University of Tokyo.

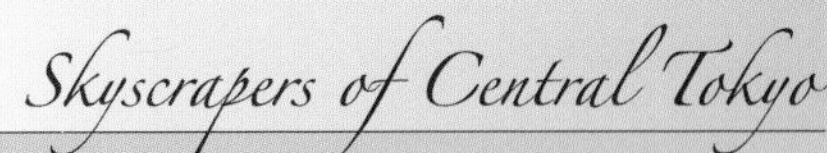

Center for Spatial Information Science, University of Tokyo

Kashiwa, Chiba Prefecture, Japan

By Akiko Takahashi and Hideto Satoh

Contact

Akiko Takahashi, Hideto Satoh

akuri@csis.u-tokyo.ac.jp, hideto@csis.u-tokyo.ac.jp

Software

ArcGIS Desktop 9, ArcGIS 3D Analyst

Printer

Canon W7200

Data Source(s)

Sinfonica Japan; Ministry of Land Infrastructure and Transport; Geographical Survey Institute; Zenrin Co., Ltd.

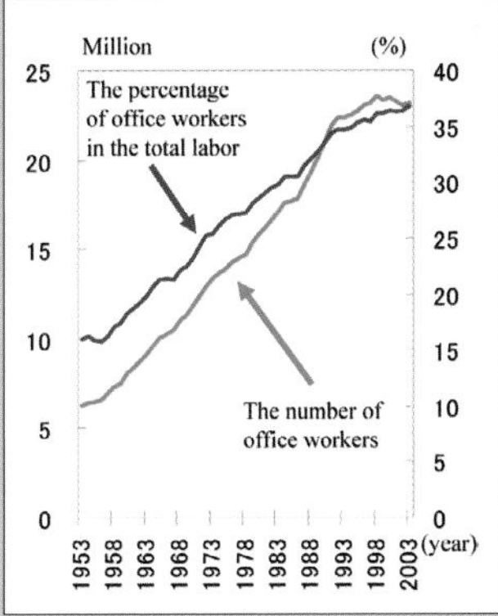

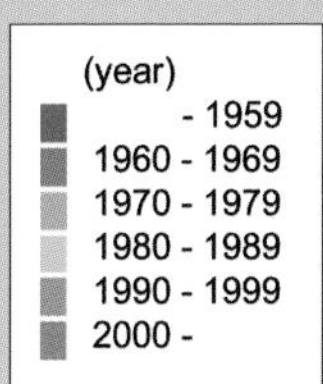

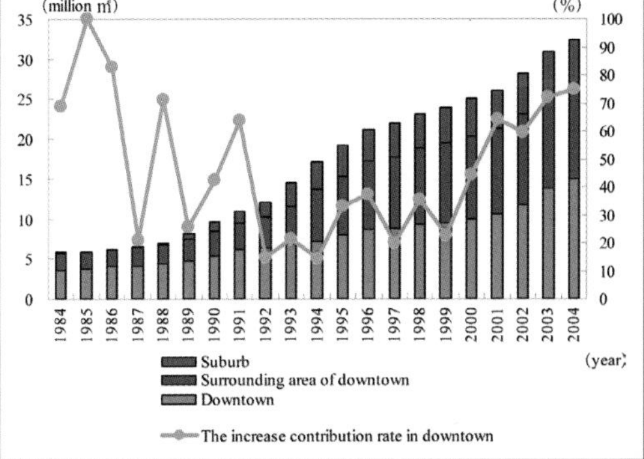

This figure indicates the locations of large office buildings by year. A majority of these buildings during and before 1989 were constructed inside the JR Yamanote Line (Tokyo Loop Line), particularly in the existing central business district such as Marunouchi and Ote-machi. On the other hand, the period from 1990 to 1999 saw the reverse of this trend. More buildings were constructed outside the same line. New office buildings were constructed in areas that did not traditionally have offices located there, including the new Tokyo waterfront subcenter.

Courtesy of Center for Spatial Information Science, University of Tokyo.

City of Phoenix

Phoenix, Arizona, USA

By Kelly Walker and Alicja Boryczko

Contact

Kelly Walker

kelly.p.walker@phoenix.gov

Software

ArcGIS Desktop 9.1

Printer

HP Designjet 1055CM

Data Source(s)

Maricopa County Assessor, City of Phoenix

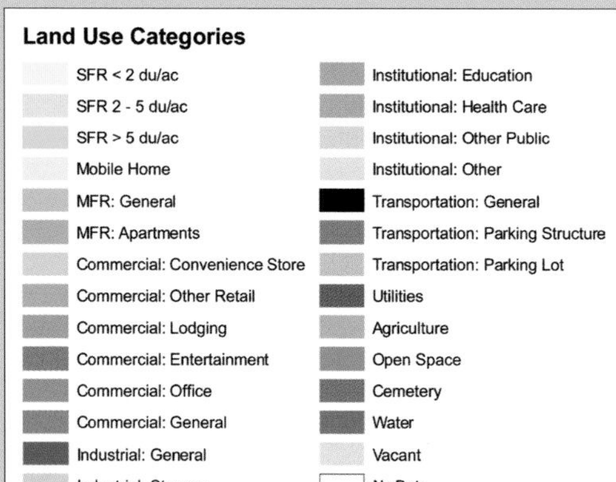

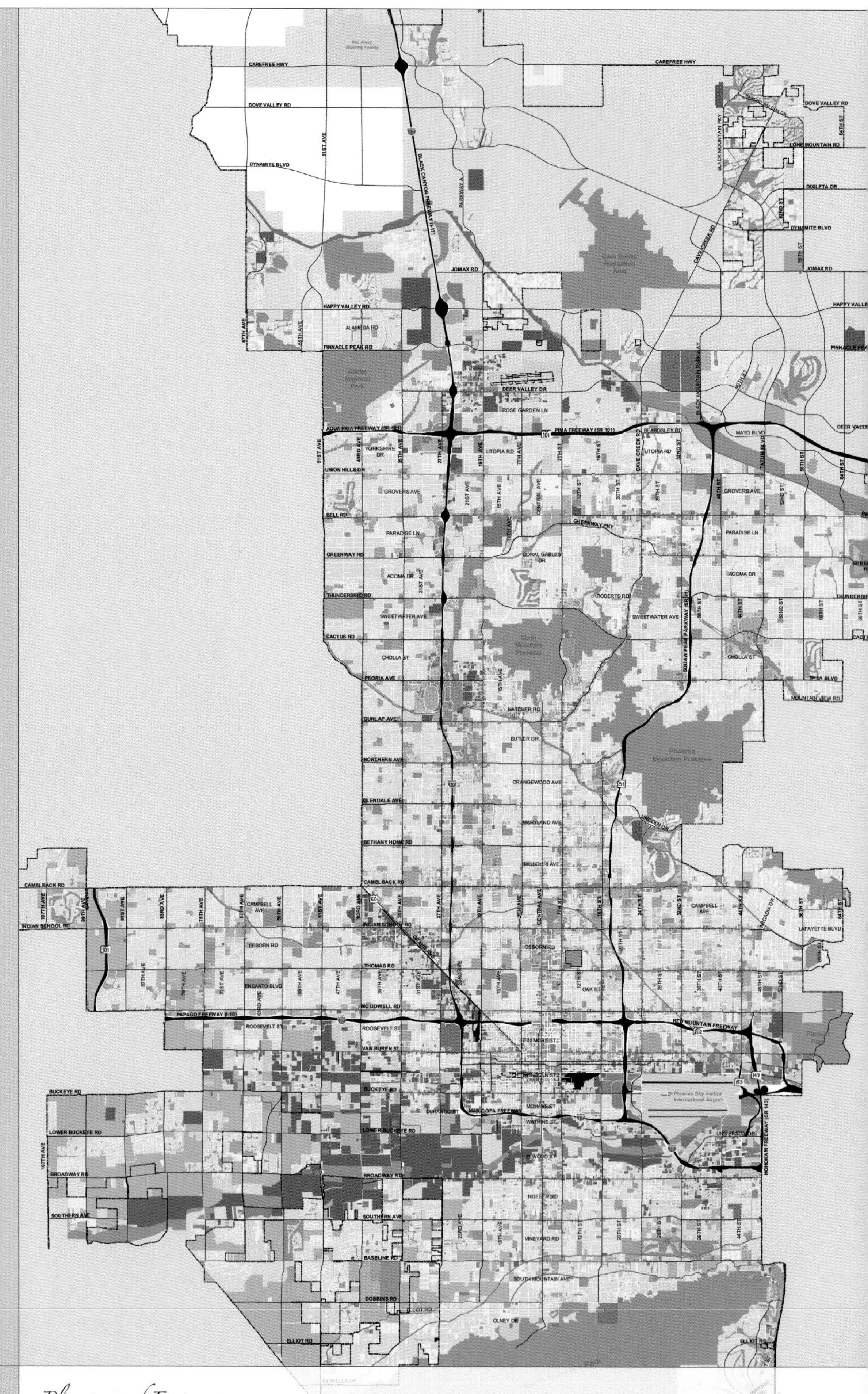

Land use is totally dependent upon physical features. But with more than 500,000 parcels in Phoenix, Arizona, totaling 512 square miles, it is difficult with a small staff to maintain a current map. To solve this dilemma, the Planning Department utilized the Maricopa County Assessor's parcel base and its associated tax codes (2,086 categories) to derive this land-use map. The county tax codes have been regrouped into 27 logical land uses. These simplified uses were selected for analysis purposes to correspond to the city's General Plan. Thus, the municipal government uses its county assessor inspectors as land-use personnel to maintain an up-to-date database and effectively save the City of Phoenix Planning Department many hours of aerial photo analysis.

Courtesy of the City of Phoenix, Arizona.

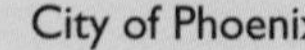

City of Phoenix

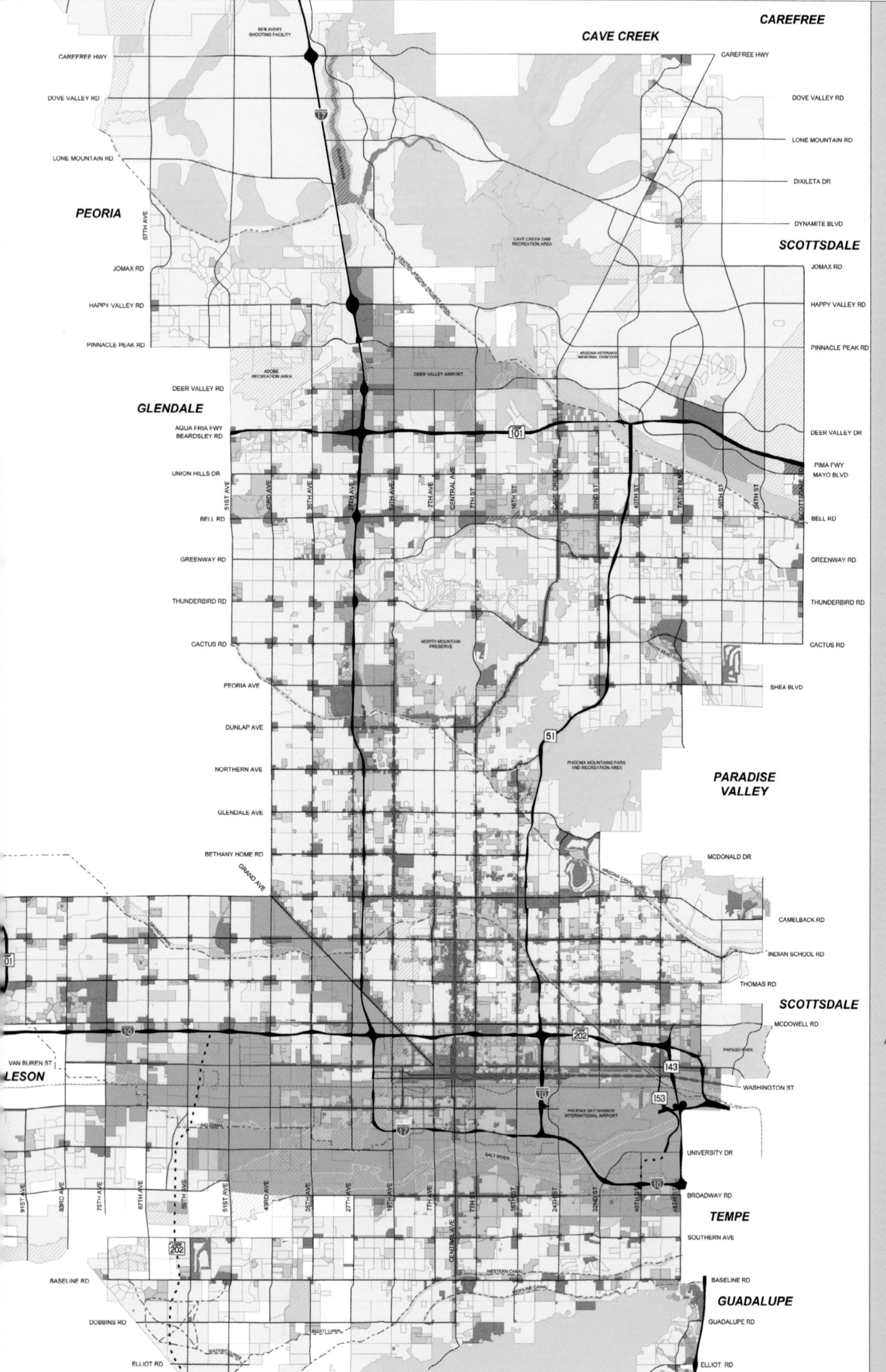

Phoenix, Arizona, USA
By Racelle Escolar

Contact
Kelly Walker
kelly.p.walker@phoenix.gov

Software
ArcGIS Desktop 9.1

Printer
HP Designjet 1055CM

Data Source(s)
Maricopa County Assessor, City of Phoenix

EXPLANATION

Generalized Zoning Classifications

- Single Family Residential
- Multi-Family Residential
- Commercial
- Industrial
- Non-Developable
- County or Other Jurisdiction

Other Features

- Arterial and Collector Streets
- Future Transportation
- Future Light Rail
- Railroad
- Canal and Watercourse
- Floodplain
- Park and/or Preserve

This zoning map for the City of Phoenix combines 53 zoning districts into five categories for viewing ease: single family, multifamily, commercial, industrial, and nondevelopable. Phoenix zoning does not recognize streets, parks, or any physical features. For example, within the more than 350 square miles zoned as "single-family residence," there are numerous physical features that consume space but do not permit housing. To provide reality to this "zoning effect," park and preserve areas have been added. In some cases these are in areas still owned by the state and are off-limits to housing, but are still zoned as residential. On the other hand, floodplain areas were added to reveal a physical characteristic that transcends zoning. Owners can still develop their land, but the potential danger from flooding is identified in these areas.

Courtesy of the City of Phoenix, Arizona.

San Diego Central Region Smart Growth Areas

San Diego Association of Governments
San Diego, California, USA
By Steve Bouchard

Contact
Steve Bouchard
sbo@sandag.org

Software
ArcGIS Desktop

Printer
HP Designjet 800ps

Data Source(s)
SANDAG GIS database

Over the past year, the San Diego Association of Governments (SANDAG), and local planning and community development directors, have been working together to identify an initial list of smart-growth areas for inclusion in the San Diego Region's Smart Growth Concept Map draft.

Smart growth is a compact, efficient, livable, and environmentally sensitive urban development pattern. It focuses future growth and infill development close to jobs, services, and public facilities to maximize the use of existing infrastructure and preserve open space and natural resources.

This map and three others (North County, North City, and South County) were used as a first step in helping local planning directors identify areas in their jurisdictions that exemplified the characteristics of the six smart-growth place types contained in SANDAG's Regional Comprehensive Plan (RCP). The planning directors also were asked whether the areas should be classified as "existing/planned" smart-growth areas or "potential" smart-growth areas based on whether they met the residential and employment targets previously defined, and whether they met certain transit service characteristics contained in the RCP.

"Existing/planned" smart-growth areas are locations where existing or planned development is consistent with RCP targets and served by existing or planned high-frequency transit. "Potential" smart-growth areas are locations with opportunities if local land-use or regional transportation plans are changed to meet targets contained in the RCP. Once the planning directors identified the areas and estimated their status (existing/planned or potential), SANDAG staff labeled the areas (e.g., 'OB 1'). They also initiated a verification process to determine whether the areas met residential and employment targets, as well as associated transit service levels. The verification process is based on data included in SANDAG's GIS. This information is updated on a periodic basis as part of the regional growth forecast update.

Courtesy of San Diego Association of Governments.

Metropolitan Transportation Commission
Oakland, California, USA
By Garlynn G. Woodsong

Contact
Garlynn G. Woodsong
gwoodsong@mtc.ca.gov

Software
ArcGIS Desktop 9.1

Printer
HP Designjet 5000ps

Data Source(s)
U.S. Census 2000 Block PL-94 Dataset

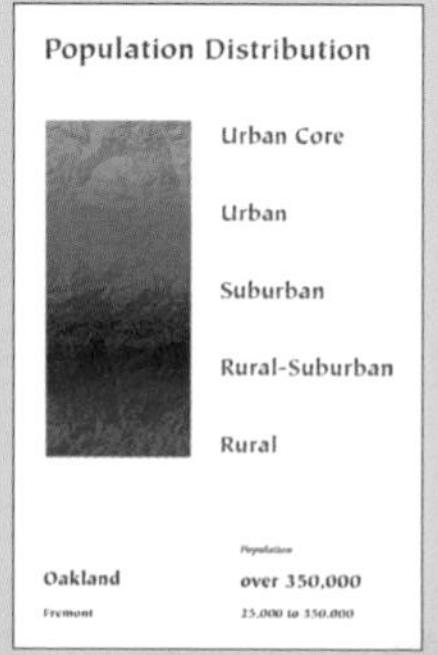

San Francisco Area

This map depicts a multidimensional surface derived from population density distribution within the nine-county San Francisco Bay Area. A hillshade grid was layered beneath the grid depicting population density from which it was derived. In effect, this creates a shaded relief image of the Bay Area that is derived from population distribution, rather than topography. As depicted in ArcScene, the inset maps show the same dataset, with heights obtained from the population field. A 0.25 factor was used to account for the extreme heights of these data values in relation to the size of the Bay Area landscape.

Courtesy of Garlynn G. Woodsong, Metropolitan Transportation Commission.

Food Insecurity and Vulnerability in Sekhukhune, Republic of South Africa

National Department of Agriculture
Pretoria, South Africa
By Johan Duvenhage

Contact
Rudolf De Munnik
rudolf@gims.com

Software
ArcGIS Desktop 9

Printer
HP Designjet 750C Plus

Data Source(s)
ArcSDE 9 on Informix

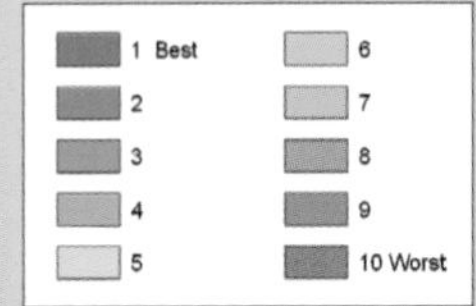

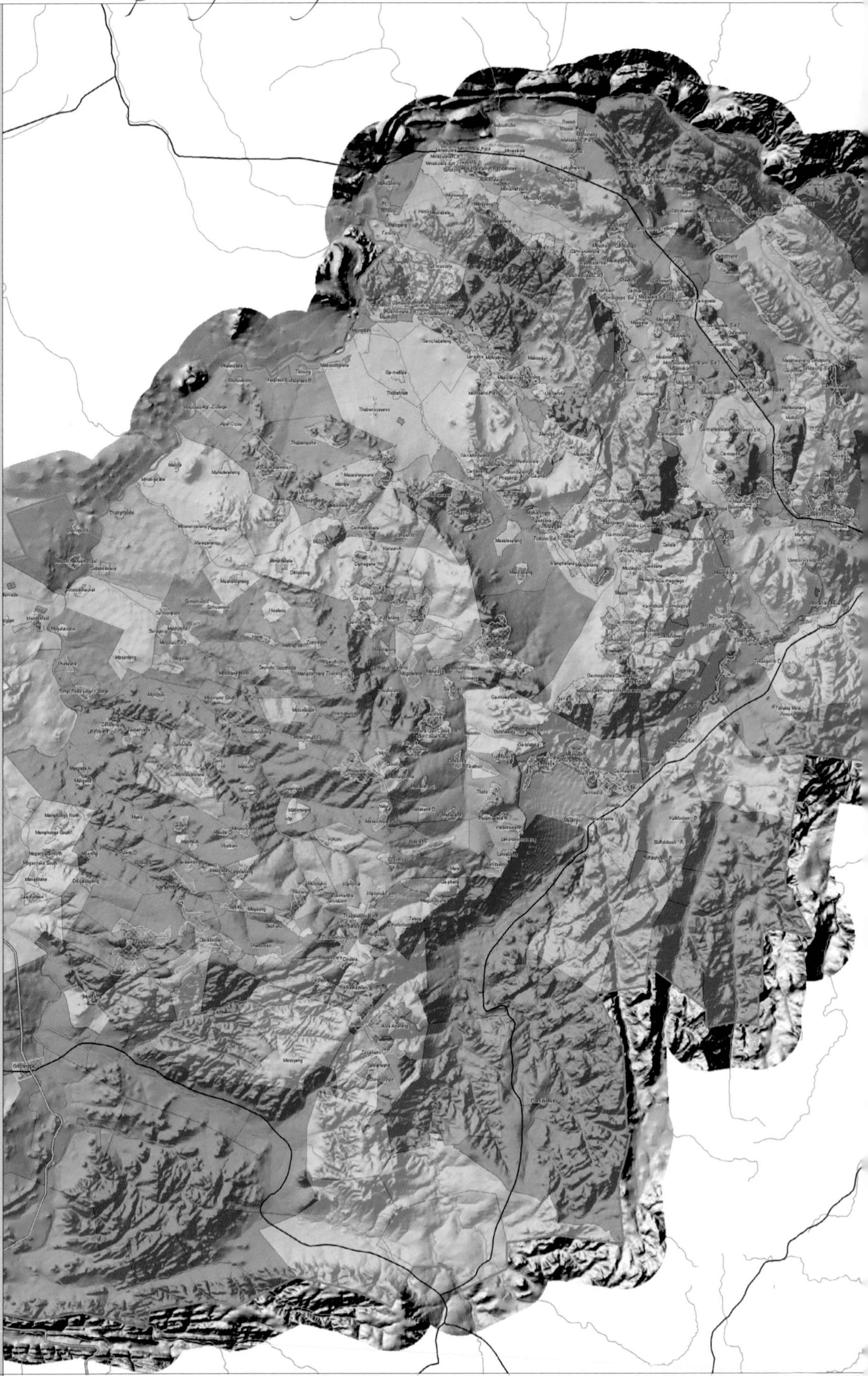

The Directorate for Food Security of the Department of Agriculture, South Africa, has set specific objectives to develop, implement, and conduct annual food security policy reviews and to establish and maintain an effective Food Insecurity and Vulnerability Information Mapping System (FIVIMS). FIVIMS is a management and decision-making support tool that provides updated information about people's status and experiences regarding hunger and vulnerability to hunger. This includes understanding people's access to food, either through their own production or purchases, as well as their health status.

In 2003 the Department of Agriculture, through a consortium of six institutions, identified a pilot FIVIMS program in the Sekhukhune Integrated Sustainable Rural Development Programme (ISRDP) node. The information produced is expected to improve the design of integrated food security interventions and also guide decision makers in allocating resources to the hungriest, poorest, and most vulnerable people and areas in South Africa. The FIVIMS process systematically assembles, analyzes, and disseminates information about people who are hungry or at risk of famine. FIVIMS also analyzes the underlying causes of problems identified in agriculture, income, markets, health, nutrition, water, or sanitation that could lead to high malnutrition rates. This map ranks 10 food security classes from "best" in green to "worst" in red within the study area of Sekhukhune.

National Department of Agriculture
Pretoria, South Africa
By Johan Duvenhage

Contact
Rudolf De Munnik
rudolf@gims.com

Software
ArcGIS Desktop 9

Printer
HP Designjet 750C Plus

Data Source(s)
ArcSDE 9 on Informix

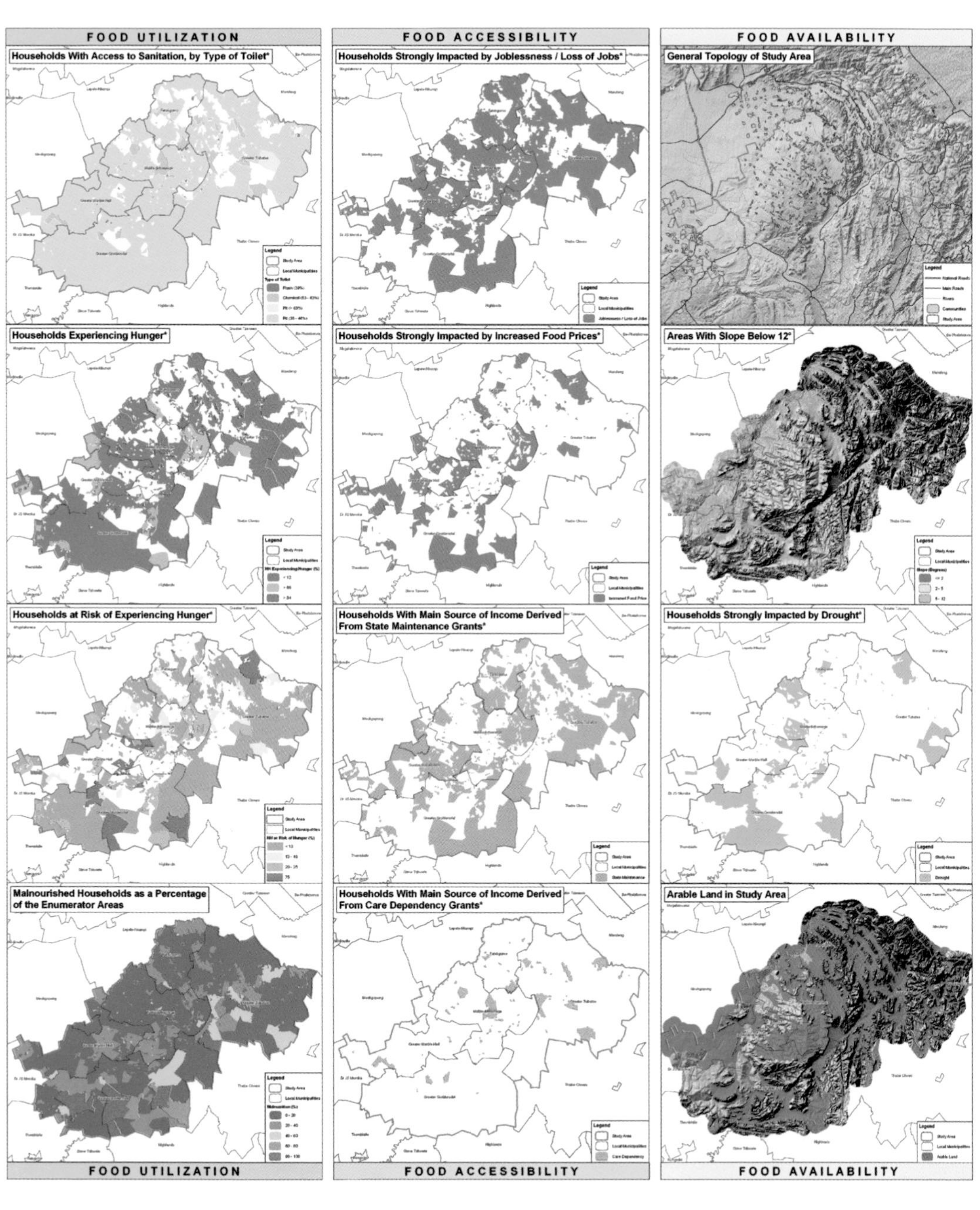

These maps identify three main drivers in Sekhukhune: food availability, food accessibility, and food utilization. Within these key drivers, drought, arable land, joblessness, food prices, grants, sanitation, hunger, and malnourishment play key roles. After the successful completion of the FIVIMS pilot phase in May 2005, the Food Security Directorate considered its next steps toward field testing and implementation of the FIVIMS in South Africa.

Courtesy of GIMS (Pty) Ltd.

Global Soil Regions

USDA Natural Resources Conservation Service
Washington, D.C., USA
Paul Reich and Hari Eswaran

Contact
Paul Reich
paul.reich@wdc.usda.gov

Software
ArcGIS Desktop 9.1

Printer
HP Designjet 1055cm

Data Source(s)
USDA-NRCS and FAO/UNESCO

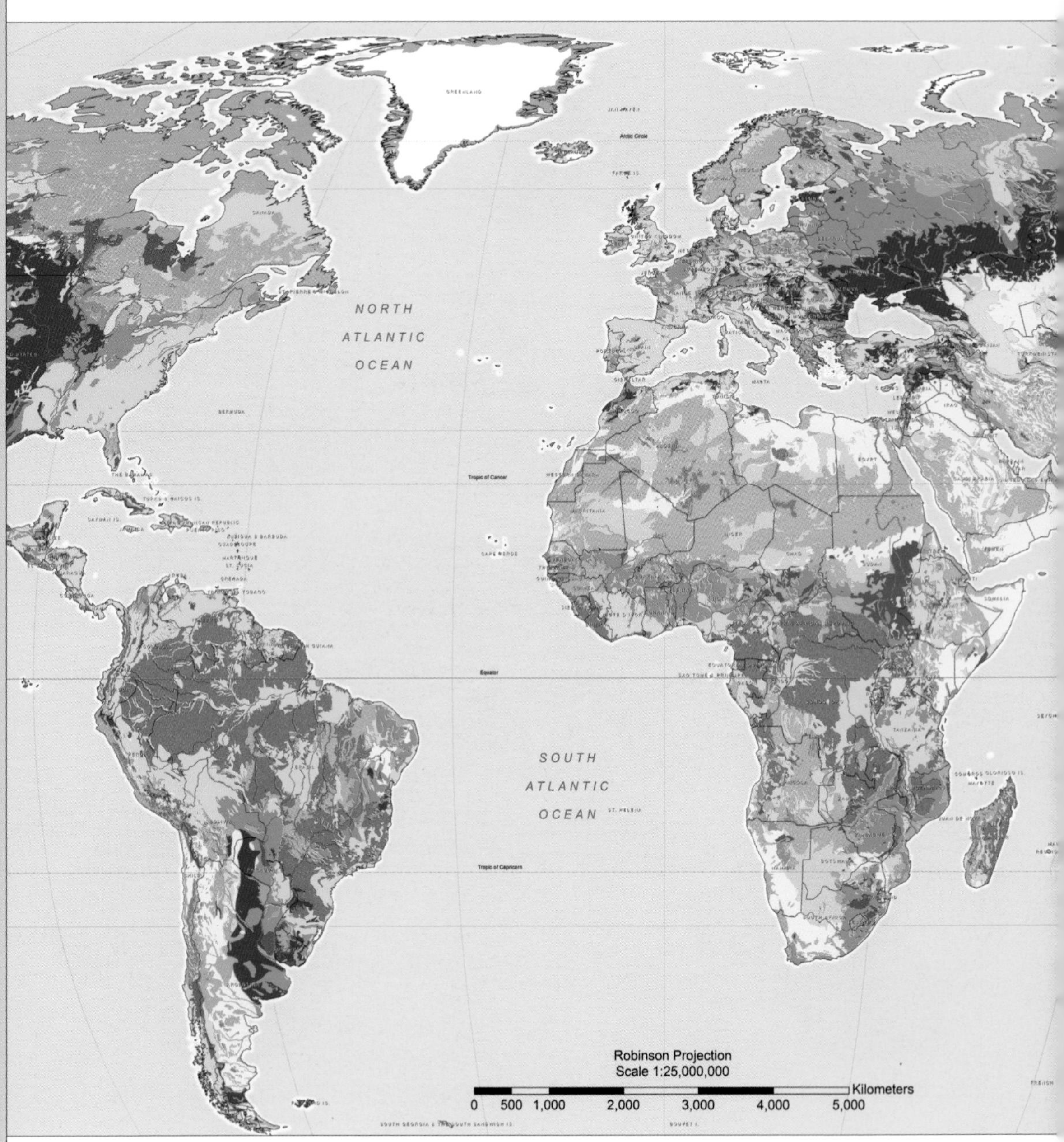

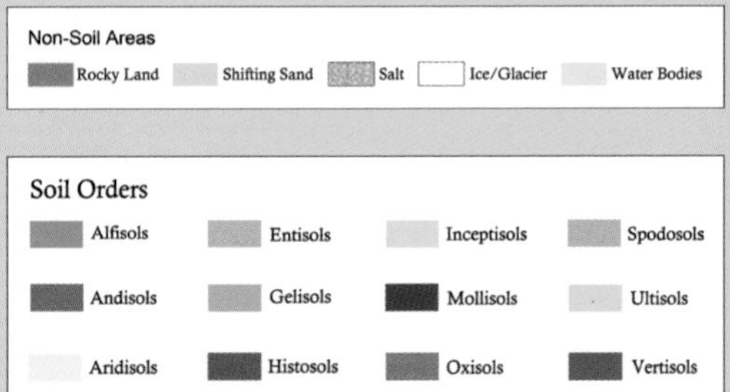

NORTH ATLANTIC OCEAN

SOUTH ATLANTIC OCEAN

Robinson Projection
Scale 1:25,000,000

0 500 1,000 2,000 3,000 4,000 5,000 Kilometers

There are tens of thousands of different kinds of soils throughout the world. Soils are classified into groups according to physical, morphological, and chemical properties. In soil taxonomy, soils are grouped into several categories, the highest being soil orders. This map shows the geographic distribution of the twelve soil orders. It has been published in the second edition of *Soil Taxonomy: A Basic System of Soil Classification for Making and Interpreting Soil Surveys*, as well as several textbooks and encyclopedias. The Global Soil Regions map is a reclassification of the Food and Agriculture Organization of the United Nations (FAO/UNESCO) Soil Map of the World to soil taxonomy suborders, and is aggregated to soil orders. As soil moisture regimes are used to define suborders, a global soil climate map was used with the FAO soil units to determine the best soil taxonomy equivalent.

Courtesy of Natural Resources Conservation Service.

City of Walla Walla

Walla Walla, Washington, USA
By Chris Owen

Contact
Chris Owen
cowen@ci.walla-walla.wa.us

Software
ArcGIS Desktop 9.1

Printer
HP Designjet 5500ps

Data Source(s)
City of Walla Walla GIS

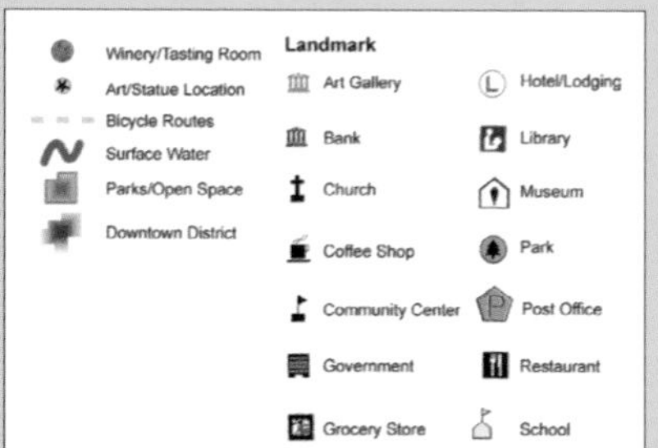

This map was created to showcase and promote the amenities and activities found in historic downtown Walla Walla, Washington, and is part of the GIS Standard Map collection maintained by the City of Walla Walla. The revitalized downtown has a variety of local amenities, including a public library, parks, restaurants, hotels, art, wine tasting, shopping, and historic architecture. The City of Walla Walla GIS Department worked with interns from Whitman College to collect and update addresses and winery and wine-tasting-room locations. Attributes collected included landmark category and landmark names. Numerous events hosted in downtown Walla Walla include the weekly farmers market, the annual Christmas parade, a bicycle criterion, and a basketball tournament. Bicycle routes, walking routes, and public transportation make getting around downtown Walla Walla easy and enjoyable.

Courtesy of the City of Walla Walla, Washington.

Regional Niagara Bicycling Map, Second Edition

The Regional Municipality of Niagara
Thorold, Ontario, Canada
By Councillor Ron Leavens, Frank Pravitz, Ken Forgeron, Tom Jamieson, John Docker, Tom Bernard, Paul Pattison, Dave Hunt, Paul Taylor, and Phil Bergen

Contact
Tom Jamieson
tom.jamieson@regional.niagara.on.ca

Software
ArcGIS Desktop 8.3, Avenza MapPublisher, Adobe Illustrator

Printer
HP Designjet 5000ps

Data Source(s)
Niagara Region GIS Services, Regional Niagara Bicycling Committee field verification

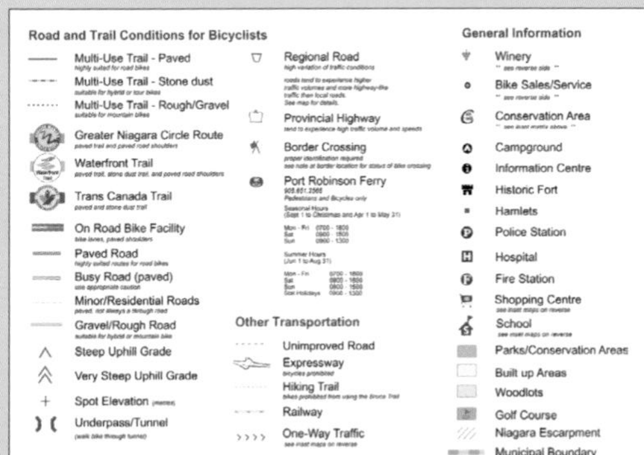

The Regional Niagara Bicycling Map (second edition) was released in February 2006. It provides comprehensive recreational, tourism, and commuter information for cyclists on roads and trails in the Canadian Region of Niagara adjacent to the United States border. The development of the map was achieved through joint contributions of regional government staff in the GIS, Planning, and Public Works departments, the Regional Niagara Bicycling Committee, and input from Freewheelers, a local bicycle touring club.

A major technical accomplishment of this map is the GIS integration. Cycling-specific data (road conditions, bike lanes, etc.) is tied to the region's Single Line Road Network (SLRN) and combined with other regional GIS layers to produce the final cartographic output. Ground truthing of cycling data was undertaken by the bicycling community. GIS integration greatly facilitates future updates and enables wider-spread use of the information (e.g., through interactive Web mapping).

Highlights of the map include an accurate and comprehensive road basemap of the Niagara Region, outlining the relative suitability of all roads for cycling, including hill locations; inset maps to assist bicyclists in navigating through larger urban areas; a broad base of local recreation and bicycle-related information (parks, bicycle operators, wineries, and other attractions); and detailed information on Niagara's famous multiuse trail systems and parkways, such as the Greater Niagara Circle Route.

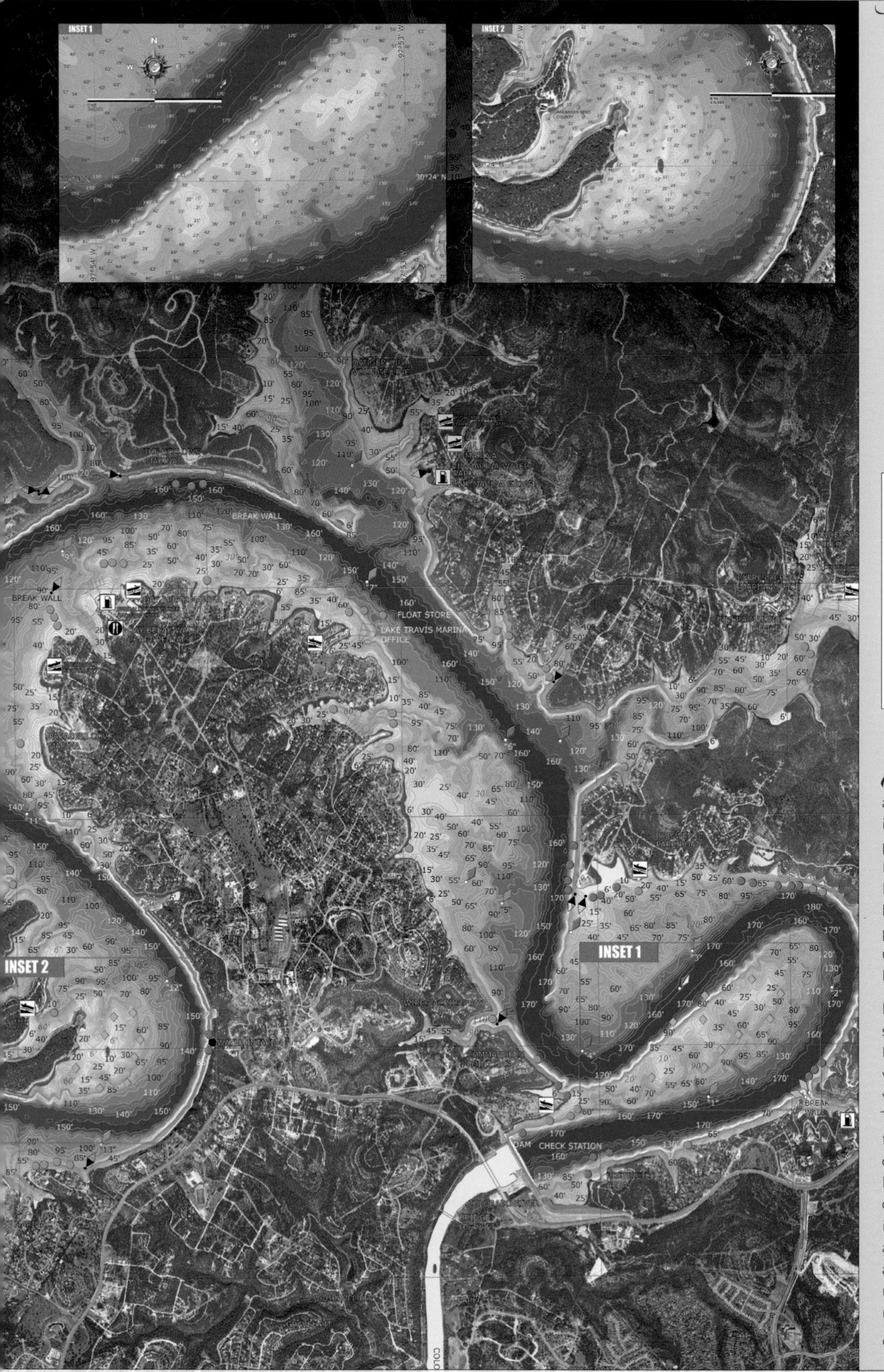

Navionics USA

Wareham, Massachusetts, USA
By Eric Sacon, Shyam Sunder Pasumarthi, Damodar Rao Vaggu, Kajrolkar Ram, and Shaun Ruge

Contact
Eric Sacon
esacon@navionics.com

Software
ArcGIS Desktop 9.1, Adobe Photoshop, Adobe Illustrator

Printer
HP Designjet 5500ps, commercial offset printer

Data Source(s)
Navionics survey, USGS

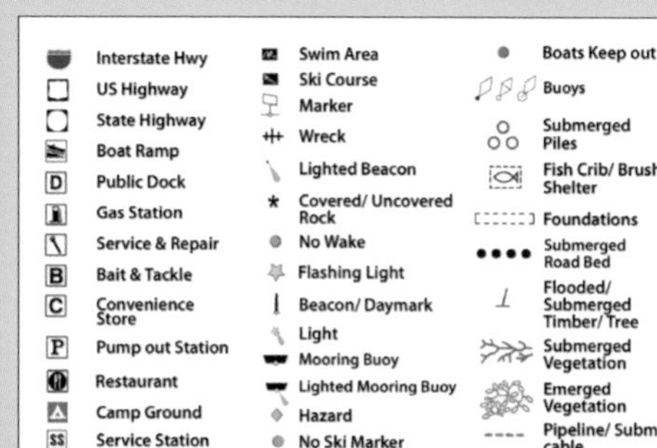

Navionics manufactures electronic navigation charts and systems for inland lakes, rivers, marine areas, and outdoor recreation. This map represents the first time in Navionics' 23-year history that a paper product has been produced. The HotMaps Premium Paper Map Series blankets more than 40 states and includes the top 300 fishing lakes in the United States. The series is part of the Navionics Bathymetry Dataset covering more than 10,000 lakes. This product is produced by seasoned anglers and intended for use by fishermen and boaters who require insight into lake-bottom structure.

Navionics' own hydrographic survey team captured the subsurface elevation data using a combination of proprietary hydrographic survey equipment and aftermarket GPS hardware and software configurations. Surveyors employ on-the-water data-editing techniques to ensure maximum accuracy. These maps have a scale range of 1:10,000 to 1:30,000 and feature one-foot bottom-contour intervals. The elevation data was interpolated in ArcGIS using custom scripts. Navigational aids, boater reference information, and points of interest were properly positioned with the help of USGS imagery and ground truthing. Bathymetric shading gradations and an extruded digital elevation model allow boaters the at-a-glance recognition of bottom features that this type of map requires. All Navionics paper maps are waterproof and tear-resistant, and they float if dropped overboard.

Courtesy of Navionics, Inc.

Summit Series — Bugaboos

Timberline Natural Resource Group

Vancouver, British Columbia, Canada
By Marcel Morin

Contact

Marcel Morin
marcel.morin@timberline.ca

Software

ArcInfo, Adobe Illustrator, Adobe Photoshop, World Construction Set

Printer

Offset press, digital direct-to-plate

Data Source(s)

Digital Forest Resource Inventory (FRI)

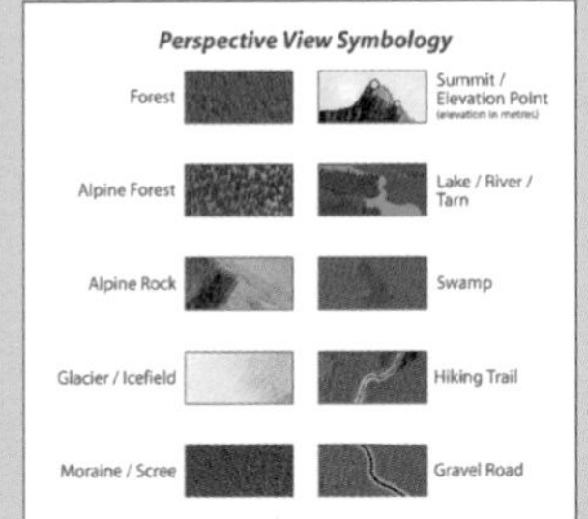

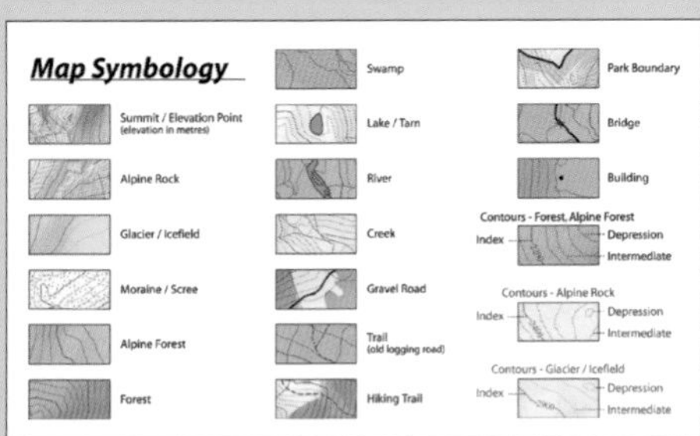

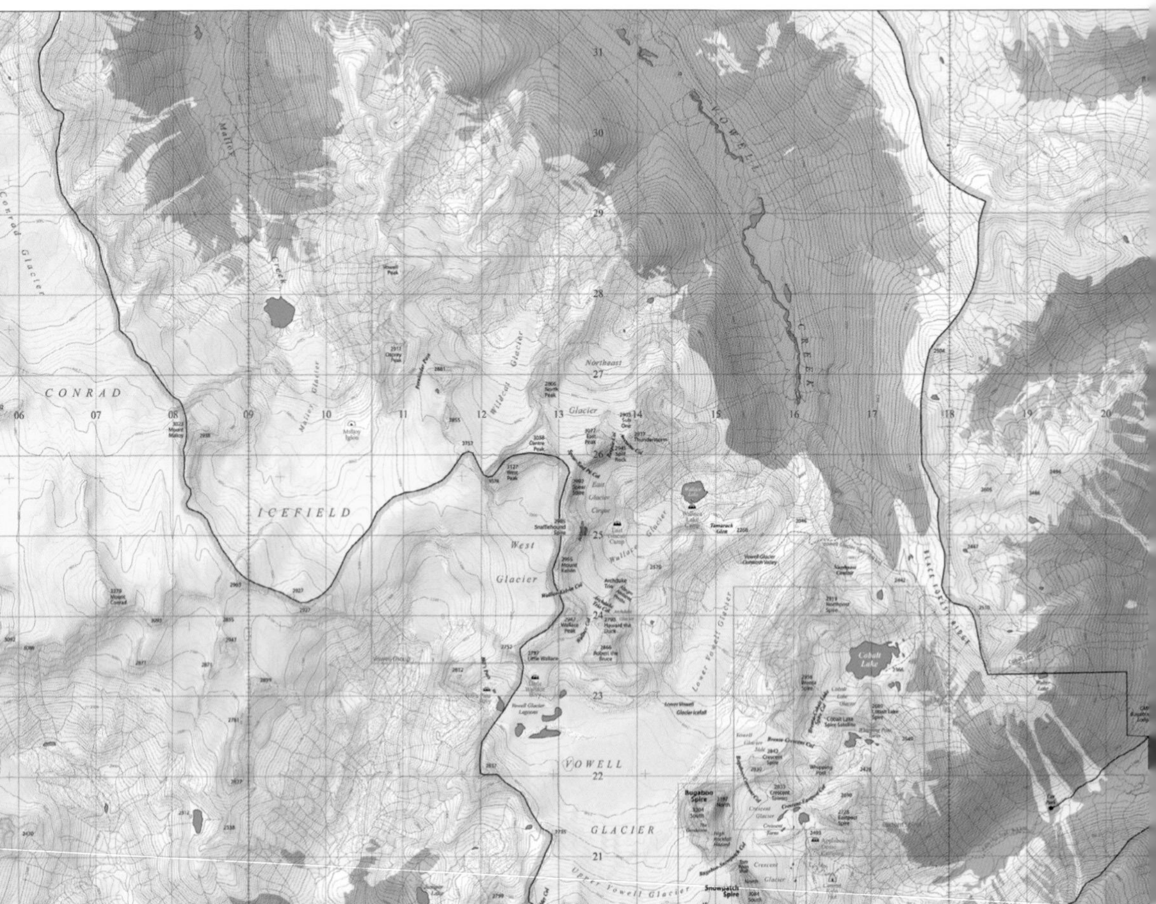

The Bugaboos are a cluster of high granite spires within Bugaboo Provincial Park in the southeast region of the province of British Columbia. The Summit Series maps were specifically created for mountaineers and printed on waterproof, tear-proof synthetic paper. This two-sided map features a realistic 3D perspective view on one side and a topo map on the other.

Courtesy of Timberline Natural Resource Group.

PB Americas, Inc.
Dallas, Texas, USA
By Runhui Liu

Contact
Miguel Otero-Jimenez, P. E.
jimenezM@pbworld.com

Software
ArcGIS Desktop, Quantm

Printer
HP Designjet 1055

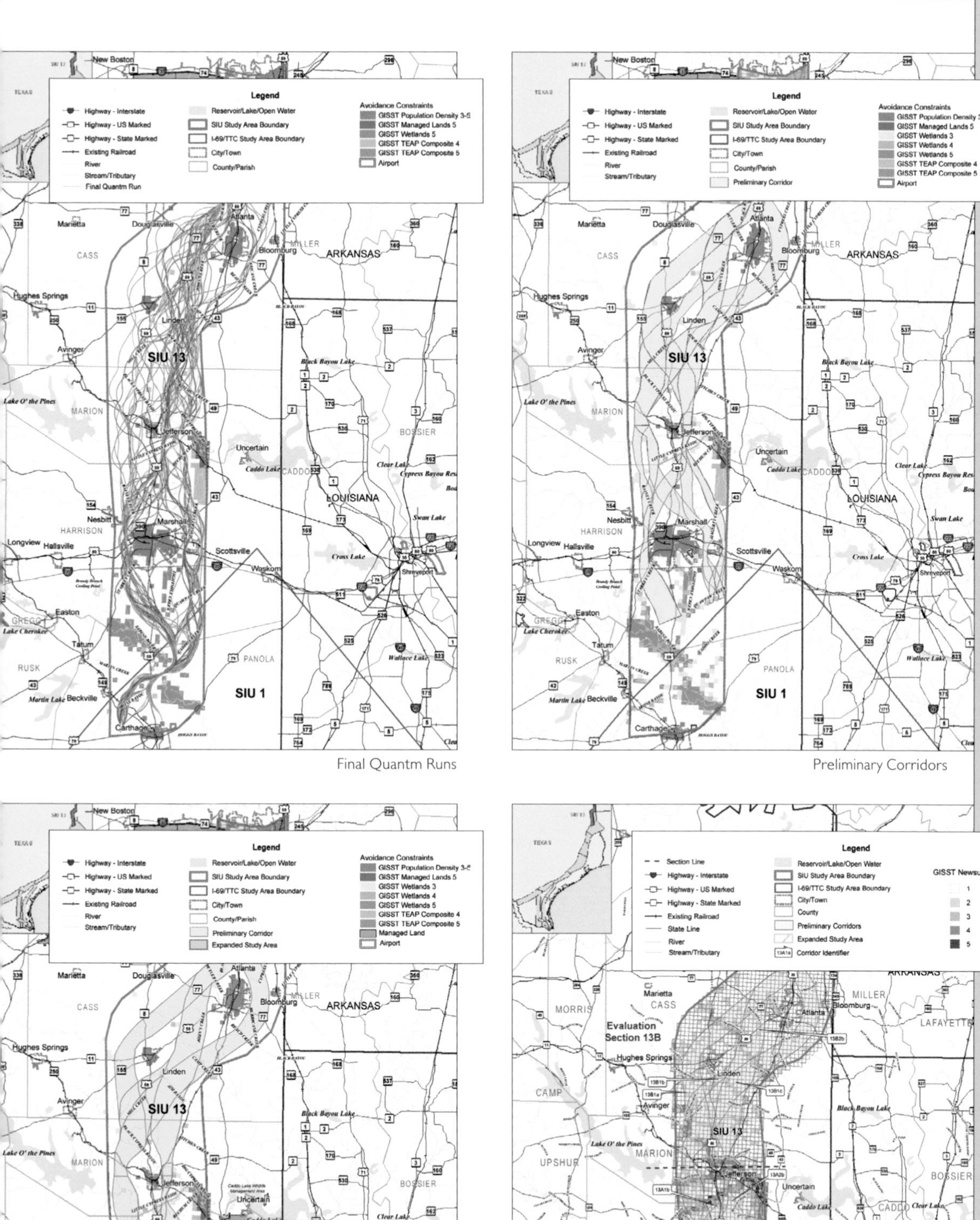

Final Quantm Runs

Preliminary Corridors

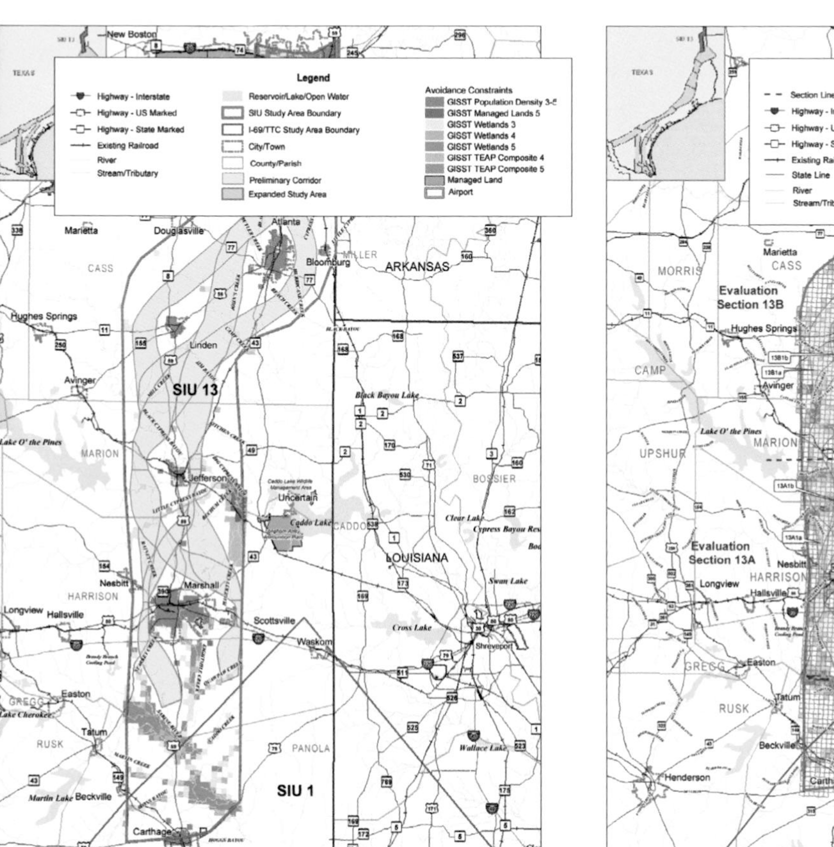

Preliminary Corridors Advanced for Evaluation

GISST Newsum

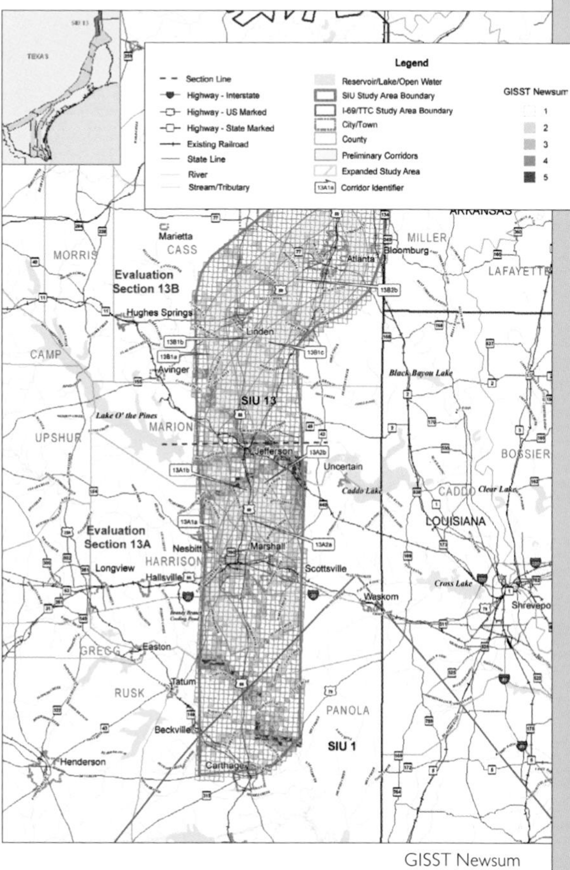

Using ArcGIS to develop, analyze, and prepare GIS data for the Quantm route optimization software, PB assisted the Texas Department of Transportation in determining the best location for the development of Interstate 69 (I-69) within the state limits. I-69 is a planned 1,600-mile national highway connecting Mexico, Canada, and the United States, with eight states involved in the project. In Texas, I-69 will be developed under the Trans-Texas Corridor master plan. The route optimization program utilizes environmental, cultural, and community GIS data combined with engineering design and costing data to evaluate and rank all possible route alternatives according to preestablished evaluation criteria.

Courtesy of PB Americas, Inc.

Visualization and Analysis of Urban Taxi Demand Survey Data

Yokohama National University
Yokohama, Kanagawa Prefecture, Japan
By Kentaro Taguchi, Satoshi Yoshida, and Satoru Sadohara

Contact
Kentaro Taguchi
kentaro@percept.ath.cx

Software
ArcGIS Desktop, ArcGIS Network Analyst, ArcGIS Spatial Analyst, ArcGIS 3D Analyst

Printer
HP Designjet 1055cm PS3

Data Source(s)
Corporate surveys, City of Sendai, Geographical Survey Institute, Ministry of Internal Affairs and Communications

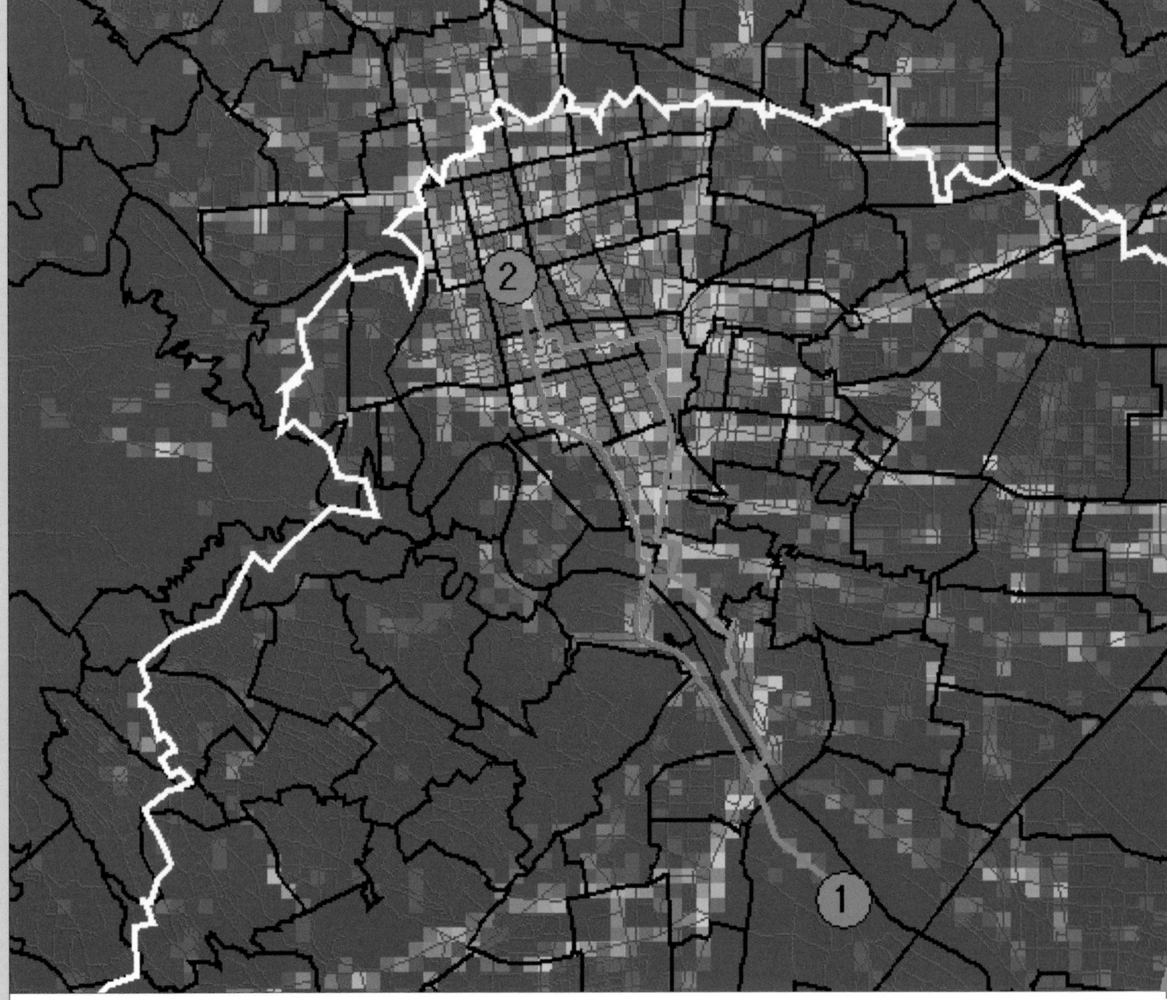

The focus of this Yokohama National University research project was to develop an efficient urban taxi management system using 417 taxis with car-mounted global positioning system (GPS) units in Sendai City, Japan. Demand estimates of taxis are dependent on time and spatial factors. Taxi demand is expressed in terms of customer population, average travel time, and purpose of travel. This is compared to regional data such as distances from train stations, building densities, and population densities. Demand is estimated for different times of the day and days of the week using regression analysis. To visualize and analyze these factors, geoprocessing tools such as kernel density, zonal statistics, and principal components were used.

The balance of supply and demand was assessed using an evaluation index. This index expresses how much time is required to reach each zone. ArcGIS Network Analyst calculates the destination of taxis in a current zone and determines an optimum route that will maximize the probability that a customer will be waiting in that zone. Not only does this create a more efficient form of dispatching, but also minimizes the amount of time wasted searching for customers and ultimately cuts down exhaust emissions.

Courtesy of Kentaro Taguchi, Satoshi Yoshida, and Satoru Sadohara, Yokohama National University.

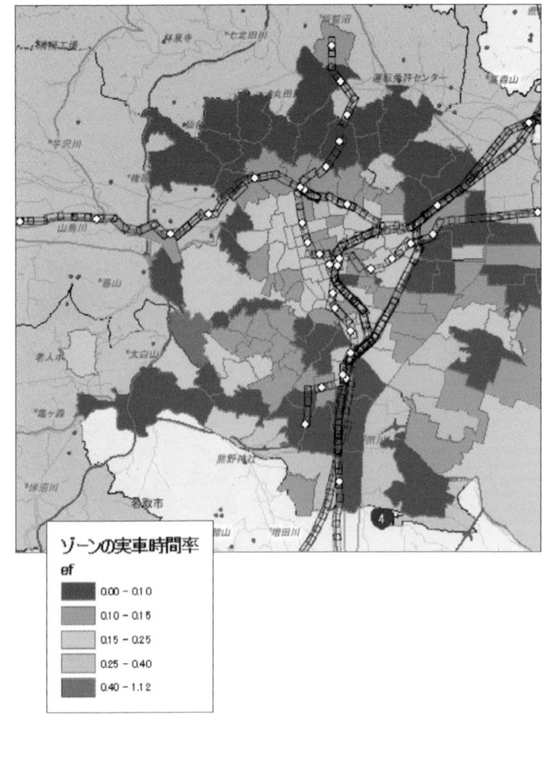

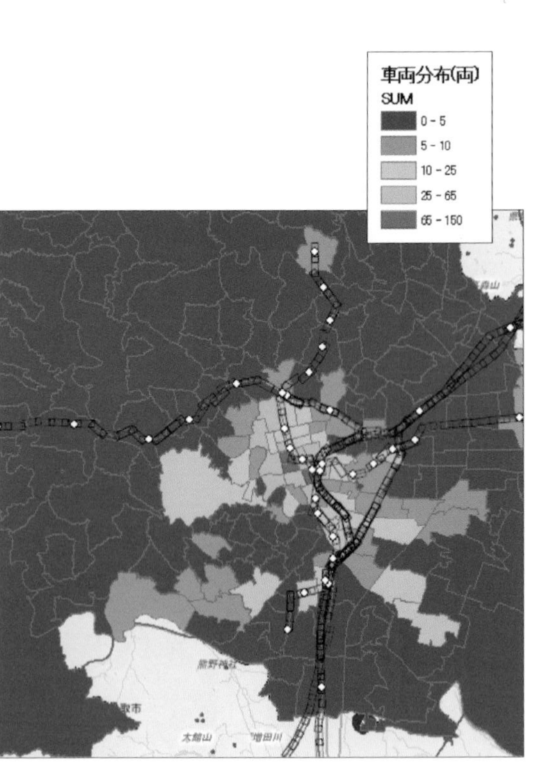

New Jersey Department of Transportation

Trenton, New Jersey, USA
By Mark Gulbinsky and Larry Chestnut

Contact

Mark Gulbinsky
mark.gulbinsky@dot.state.nj.us

Software

ArcGIS Desktop 9.1, Adobe Illustrator

Printer

HP Designjet

Data Source(s)

New Jersey Department of Transportation (NJDOT),
New Jersey Department of Environmental Protection (NJDEP),
National Oceanic and Atmospheric Administration (NOAA),
New York Department of Transportation (NYDOT),
Pennsylvania Department of Transportation (PENNDOT),
Delaware Department of Transportation (DELDOT)

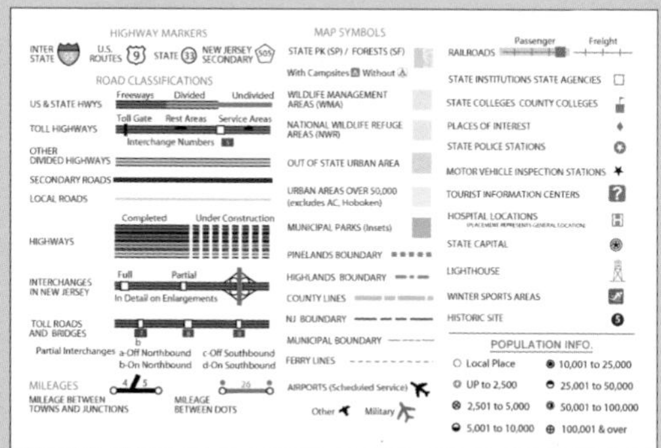

The New Jersey Department of Transportation (NJDOT) has published State Transportation Maps for more than 40 years. In 2006–2007, the state agency for the first time used ArcGIS Desktop software to create this map instead of the CADD (computer-aided design and drafting) application used previously. Using a GIS for the source data provided greater spatial accuracy, as well as more up-to-date highway geometries through connections to other NJDOT databases. Data in a variety of formats from other state departments of transportation was integrated seamlessly, with a minimum of difficulty. Mapmakers spatially created features such as the index of towns and mileage chart with increased precision.

Courtesy of New Jersey Department of Transportation.

Generalized Map Uses: Athens Metropolitan Area

Attiko Metro S.A.
Athens, Greece
By Attiko Metro S.A.

Contact
Demetrios Panayotakopoulos
dpanayotakopoulos@ametro.gr

Software
ArcGIS Desktop 9.2

Printer
HP Designjet 4000

Data Source(s)
Land Use Survey (1995–1996)

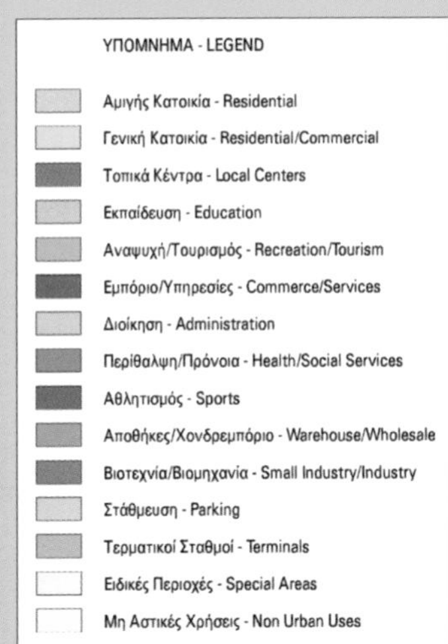

The Metro Development Study is aimed at long-term planning of the Metro System for the Athens Metropolitan Area. The Metro System—seen within the framework of an integrated public transport system that incorporates all means of continuously interacting transport, including private means—plays a significant role in the strategic plan of the study area.

A suitable transportation model was developed for strategic planning purposes. It assists in developing and testing alternative scenarios for Metro System extensions that forecast future transport demand. These are defined by socioeconomic characteristics, land-use distribution, and supply needs according to the characteristics of the public transport system and private means.

Courtesy of Attiko Metro S.A.

City of Hampton
Hampton, Virginia, USA
By Jonah Adkins

Contact
Jonah Adkins
jadkins@hampton.gov

Software
ArcGIS Desktop 9.1

Printer
HP Designjet 1055cm

Data Source(s)
Hampton ArcSDE database

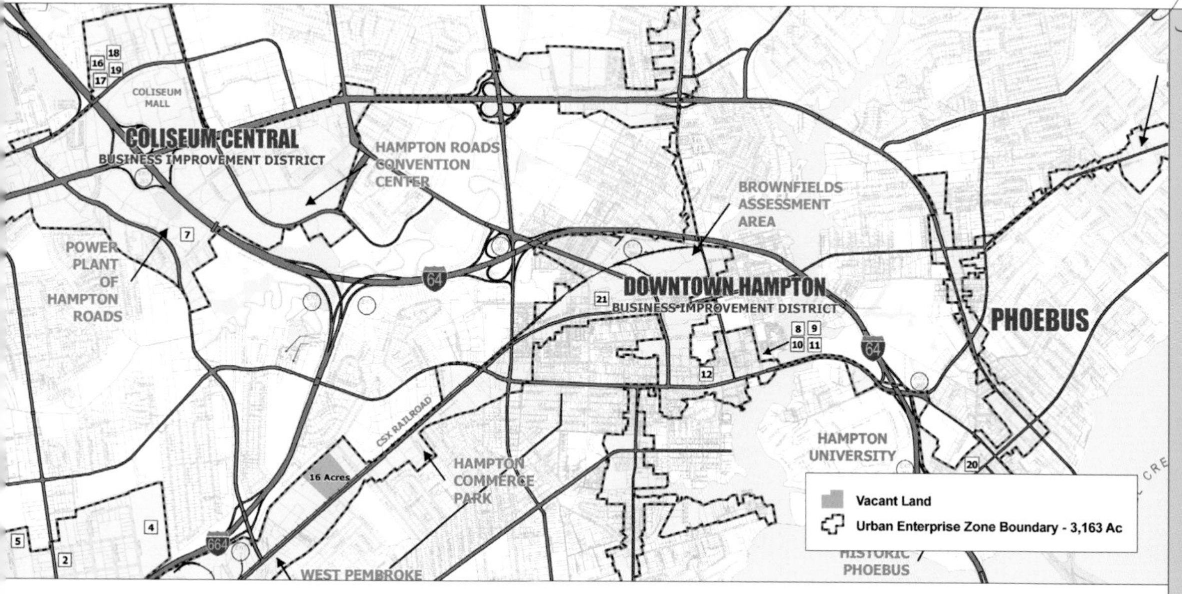

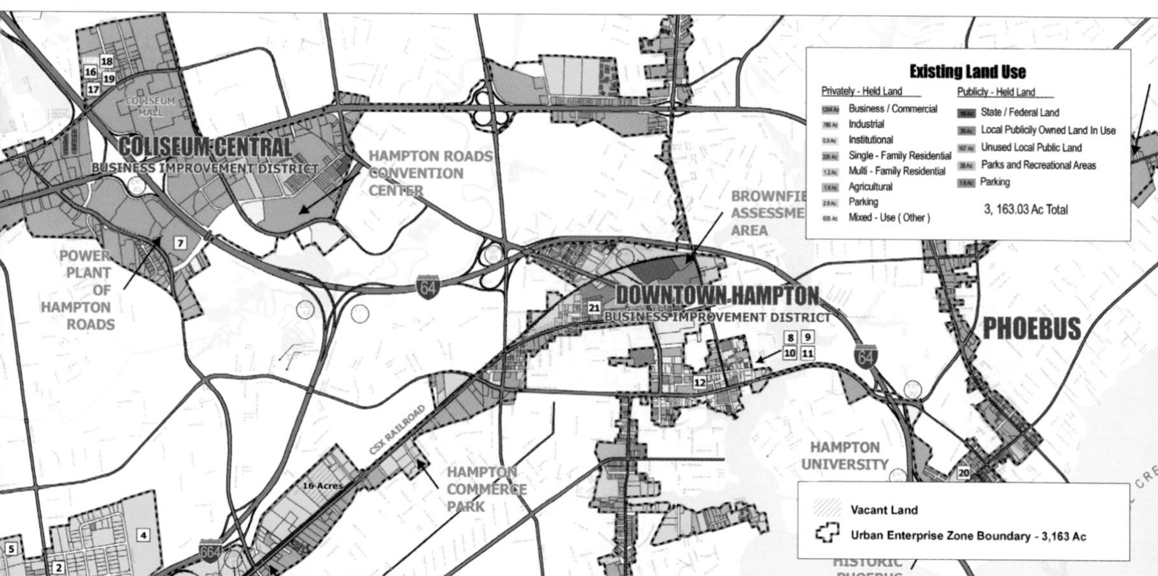

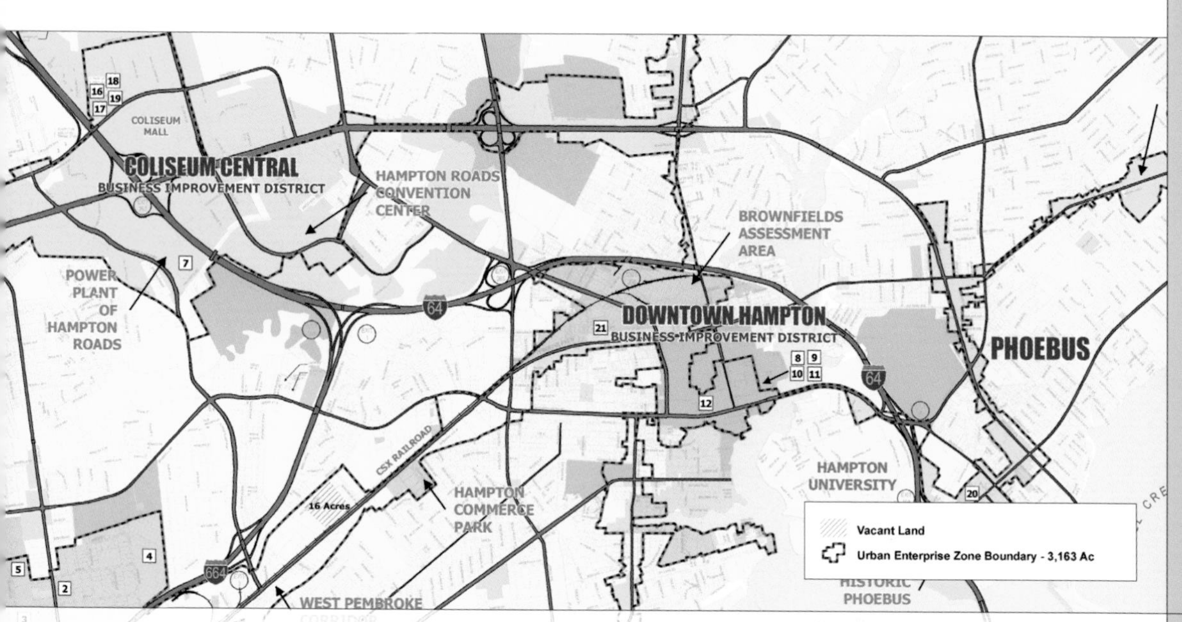

The Hampton Urban Enterprise Zone encompasses 3,440 acres and approximately one-third of the commercial property in the city. The state of Virginia defines an enterprise zone as an economically distressed, distinct geographical area of a county, city, or town. The state and local partnership seeks to improve conditions within a targeted area where incentives are provided to encourage new investment. Currently, 1.8 billion dollars of private investment has been planned and committed to the Hampton Urban Enterprise Zone, much of it focused around the Mercury Boulevard and Interstate 64 corridor.

As part of Hampton's application process for this designation, a series of three maps was created. The first map defined the proposed contiguous area, with major development areas noted. The second map highlighted the future land use for the proposed area. The third map detailed the zoning for the proposed area as well as the rest of the city. These maps played a key role in the application process and required a strong cooperative effort between Hampton's Department of Economic Development and the GIS staff.

Courtesy of Jonah Adkins, City of Hampton, Virginia.

RTI INTERNATIONAL
Research Triangle Park, North Carolina, USA
By David Chrest

Contact
David Chrest
davidc@rti.org

Software
ArcGIS Desktop 9.1, ArcGIS Spatial Analyst

Printer
HP Designjet 800ps

Data Source(s)
U.S. Census

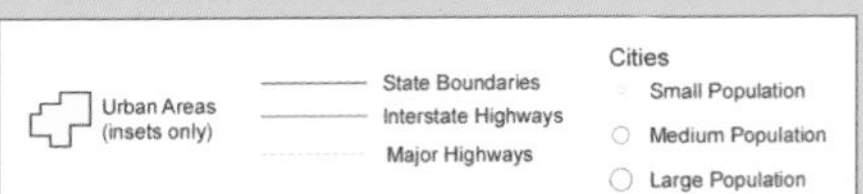

The Models of Infectious Disease Agent Study (MIDAS) is a collaboration of research and informatics groups, funded by the National Institutes of Health. Its goal is to develop computational models of the interactions between infectious agents and their hosts, the spread of disease, prediction systems, and response strategies. MIDAS models are useful not only to other scientists, but also to decision and policy makers.

As part of the informatics group, the RTI GIS program provides geographic data and has built a comprehensive geodatabase so that MIDAS researchers can take advantage of a wide variety of geographic data in constructing their models.

The maps were designed in a manner that allows users to visualize as elevations the counts of commuters arriving at Census Tracts of Work. A hillshade was created with the output from the ArcGIS Spatial Analyst Topo to Raster interpolation tool layered on top with transparency. The hillshade and commuting "elevation" rasters clipped to a detailed outline of the United States, and shading to mimic terrain maps, created an effective surface map thats allows the user to quickly identify areas and patterns of high to low movement of commuters. Sixteen urban-area insets are included on the original map with Alaska, Hawaii, and the lower 48 states. The map symbolizes state boundaries, urban areas, interstates, and major highways. Small, medium, and large cities and suburbs are also labeled. The Places Gaining People vs. Losing People map shows net change levels of inbound versus outbound commuters by census tract. Areas losing—rather than gaining—net daily commuters are symbolized as light- to dark-blue "below sea level" depths. U.S. commuting data is based on Census Spatial Tabulation: Census Tract of Work by Census Tract of Residence (STP64).

Courtesy of RTI International.

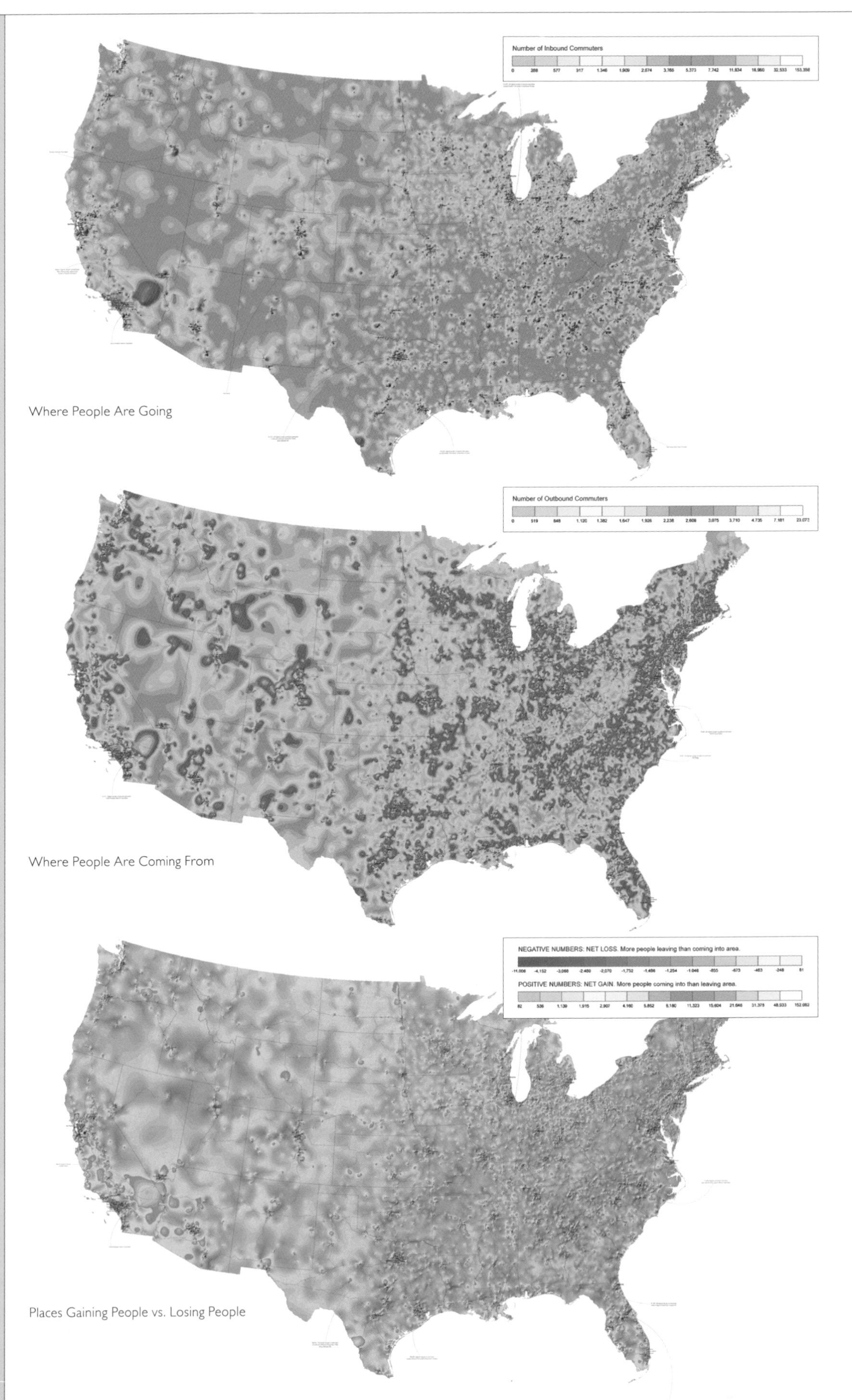

Where People Are Going

Where People Are Coming From

Places Gaining People vs. Losing People

Where People Are Going

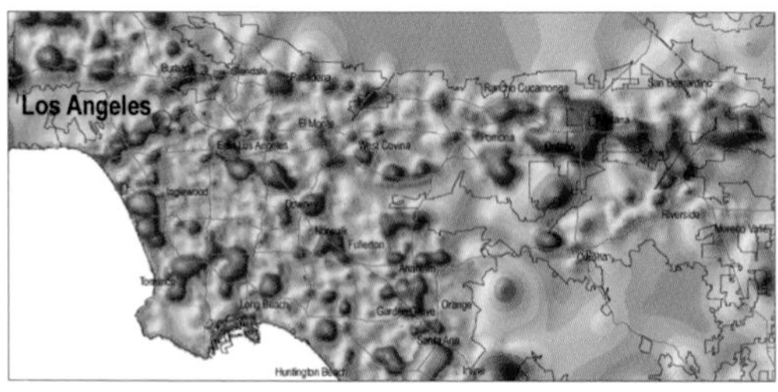

Los Angeles, California

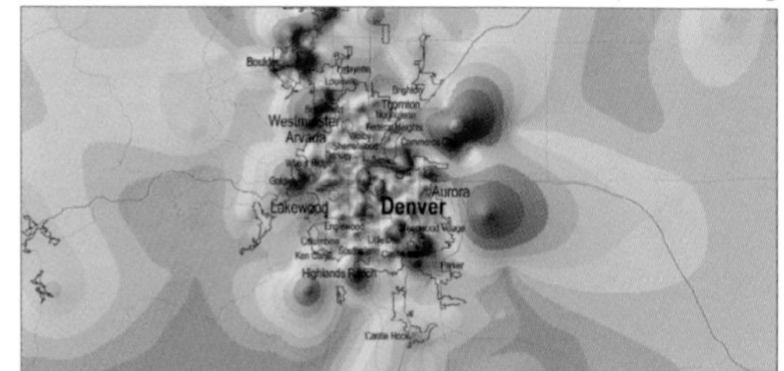

Denver, Colorado

Chicago, Illinois

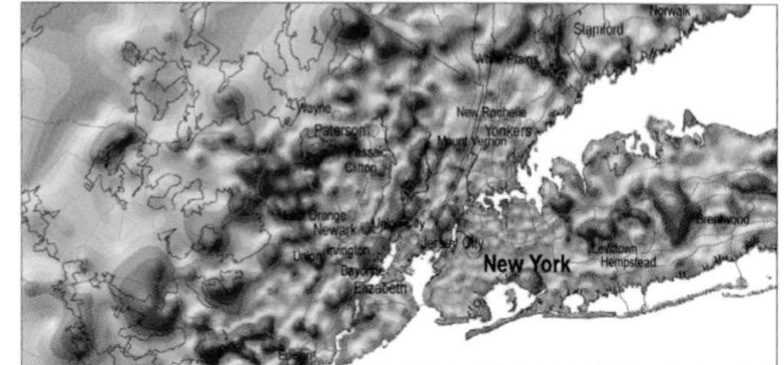

New York, New York

Where People Are Coming From

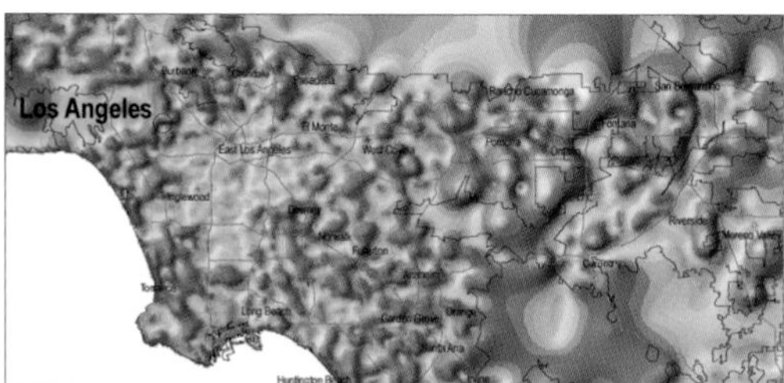

Los Angeles, California

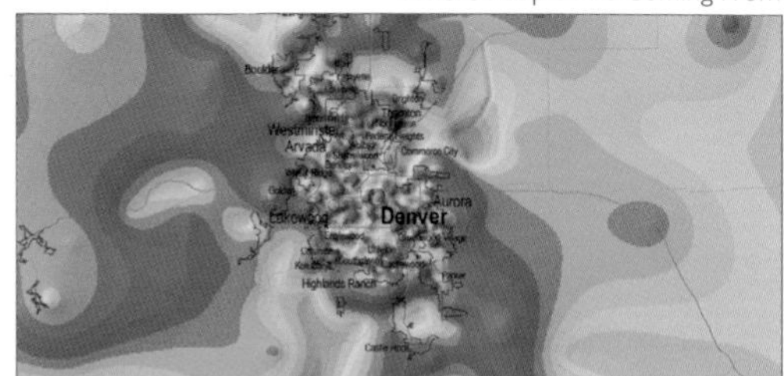

Denver, Colorado

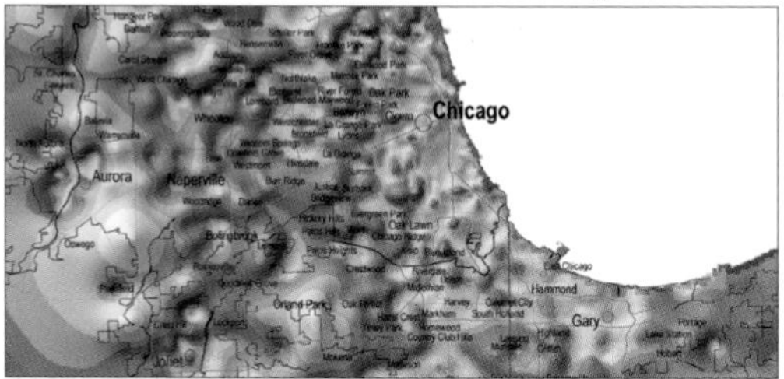

Chicago, Illinois

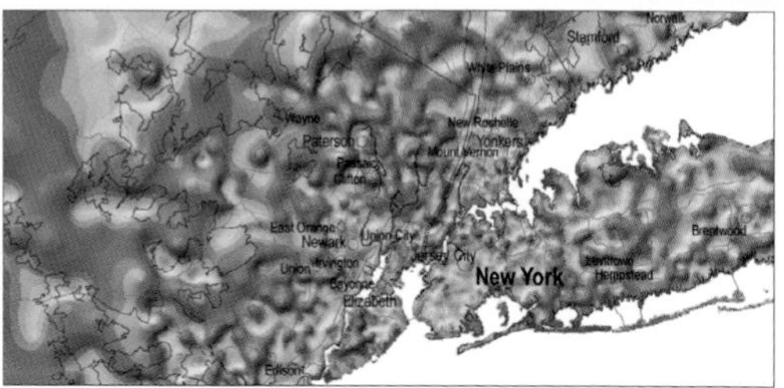

New York, New York

Places Gaining People vs. Losing People

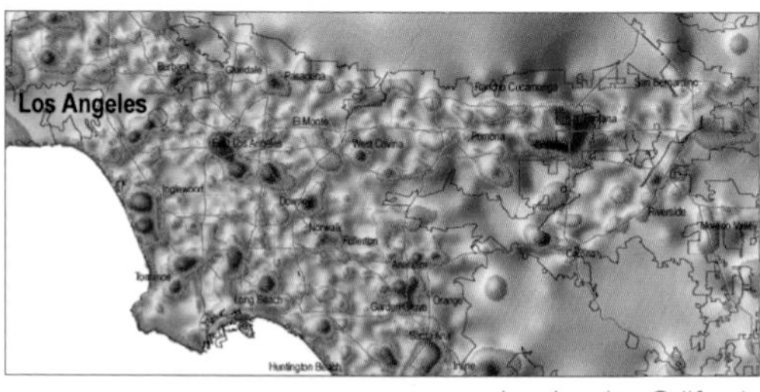

Los Angeles, California

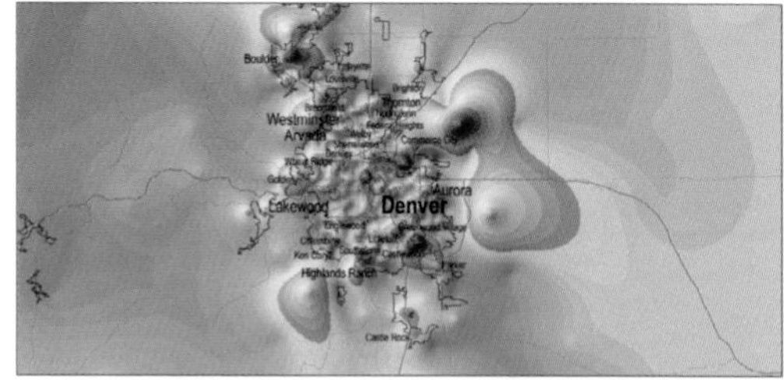

Denver, Colorado

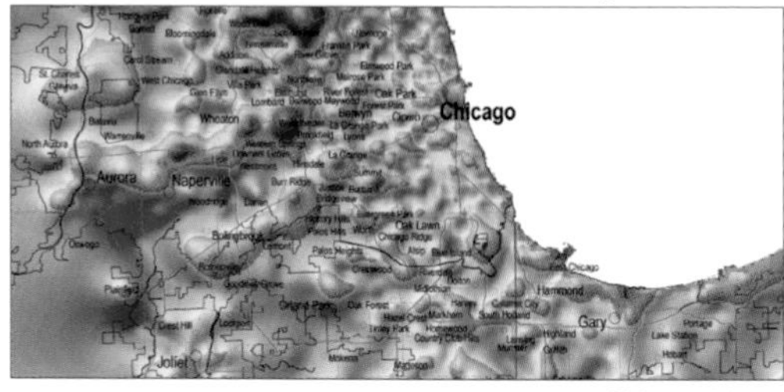

Chicago, Illinois

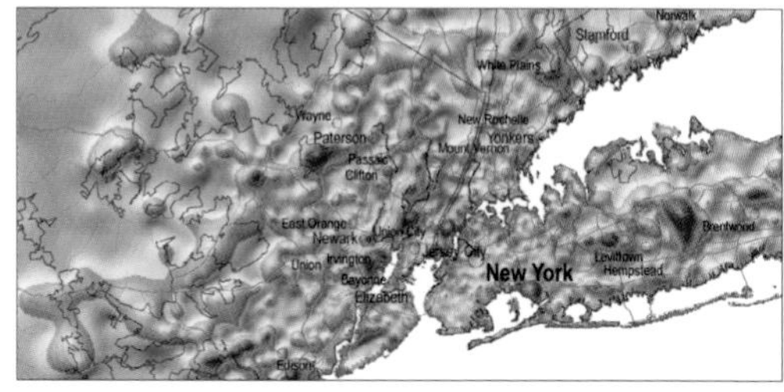

New York, New York

Lakewood Traffic Collisions

City of Lakewood

Lakewood, California, USA

Michael Mercurio, Kayla Folkins, and Michael Jenkins Jr.

Contact

Michael Jenkins Jr.

mjenkins@lakewoodcity.org

Software

ArcGIS Desktop 9.1

Printer

HP Designjet 5000ps

Data Source(s)

City of Lakewood, Rand McNally & Co., Los Angeles County Sheriff's Department

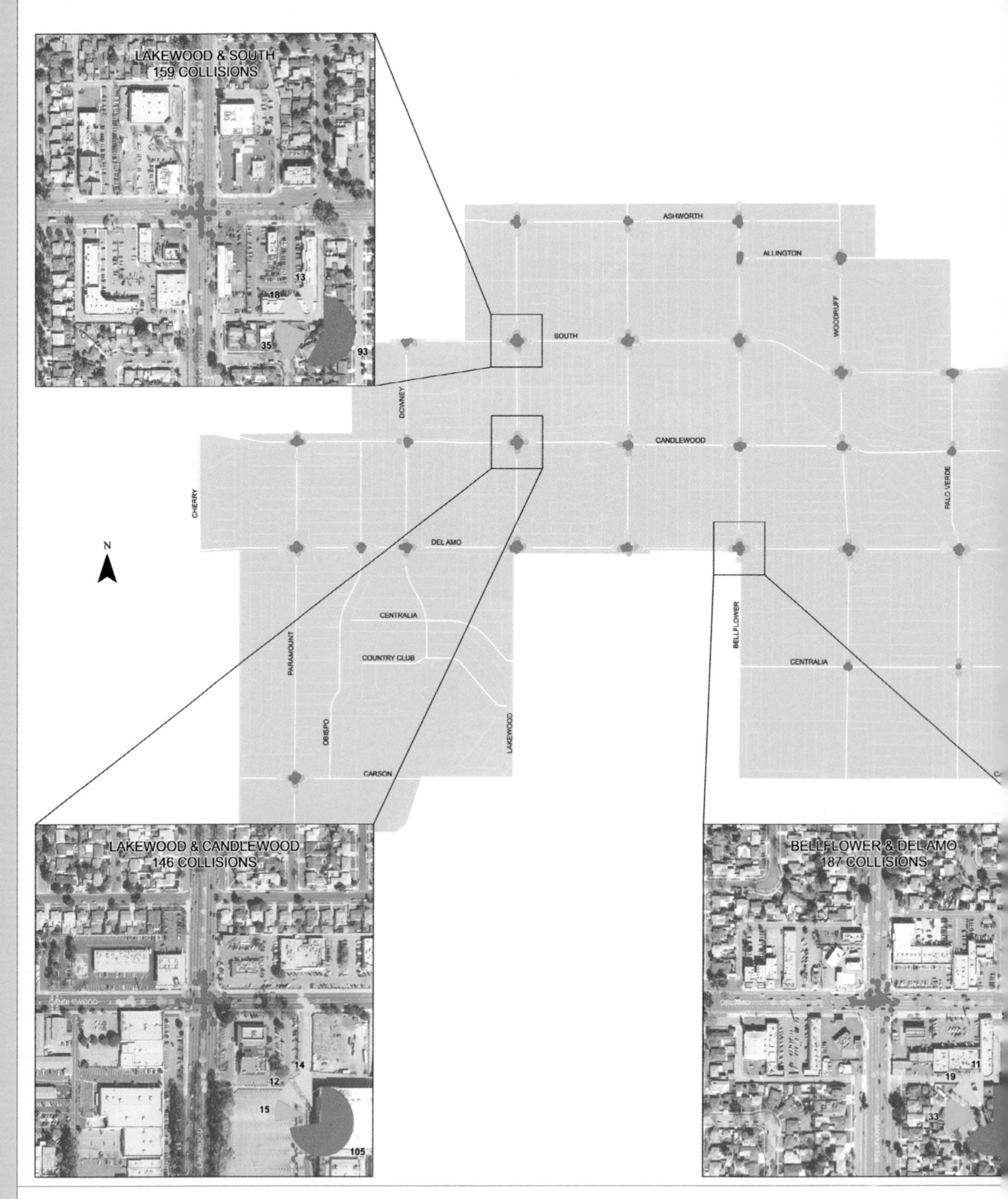

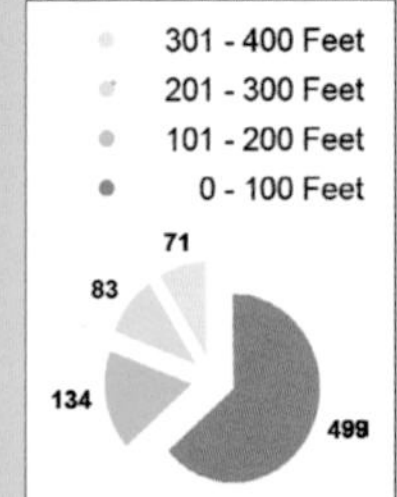

In 1999, City of Lakewood GIS staff began digitizing reported traffic collisions using data provided by the Lakewood Sheriff's Department. Along with the location of each collision, information such as the time and day, alcohol and drug influence, lighting conditions, and other factors related to the collision were captured. More than 10,000 collisions since 1997 have been digitized.

The original purpose of this map was to showcase the collision data created by—and stored in—the city's GIS. However, after viewing early versions of the map, city government staff began to wonder where the "Top 5" intersections for traffic collisions were and how the collisions were distributed within them. To find the answer, all major intersections in Lakewood were identified. From the center of each intersection, four 100-foot ringed buffers were created in succession starting from the center of each major intersection. Using these buffers, the total number of collisions occurring within 400 feet of each intersection could be determined. This technique provided Lakewood with a "Top 5" list as well as the ability to measure the density of collisions in each intersection at regular intervals.

Lakewood Infrastructure Map

City of Lakewood

Lakewood, California, USA
By Michael Mercurio, Kayla Folkins, and Michael Jenkins, Jr.

Contact

Michael Jenkins, Jr.
mjenkins@lakewoodcity.org

Software

ArcGIS Desktop 9.1

Printer

HP Designjet 5000ps

Data Source(s)

City of Lakewood, Los Angeles County Assessor

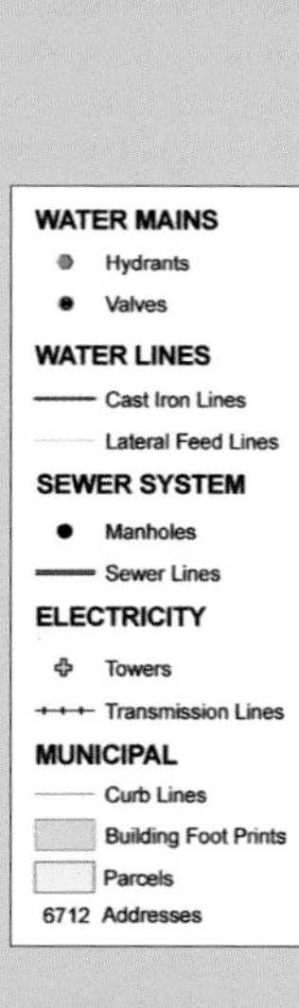

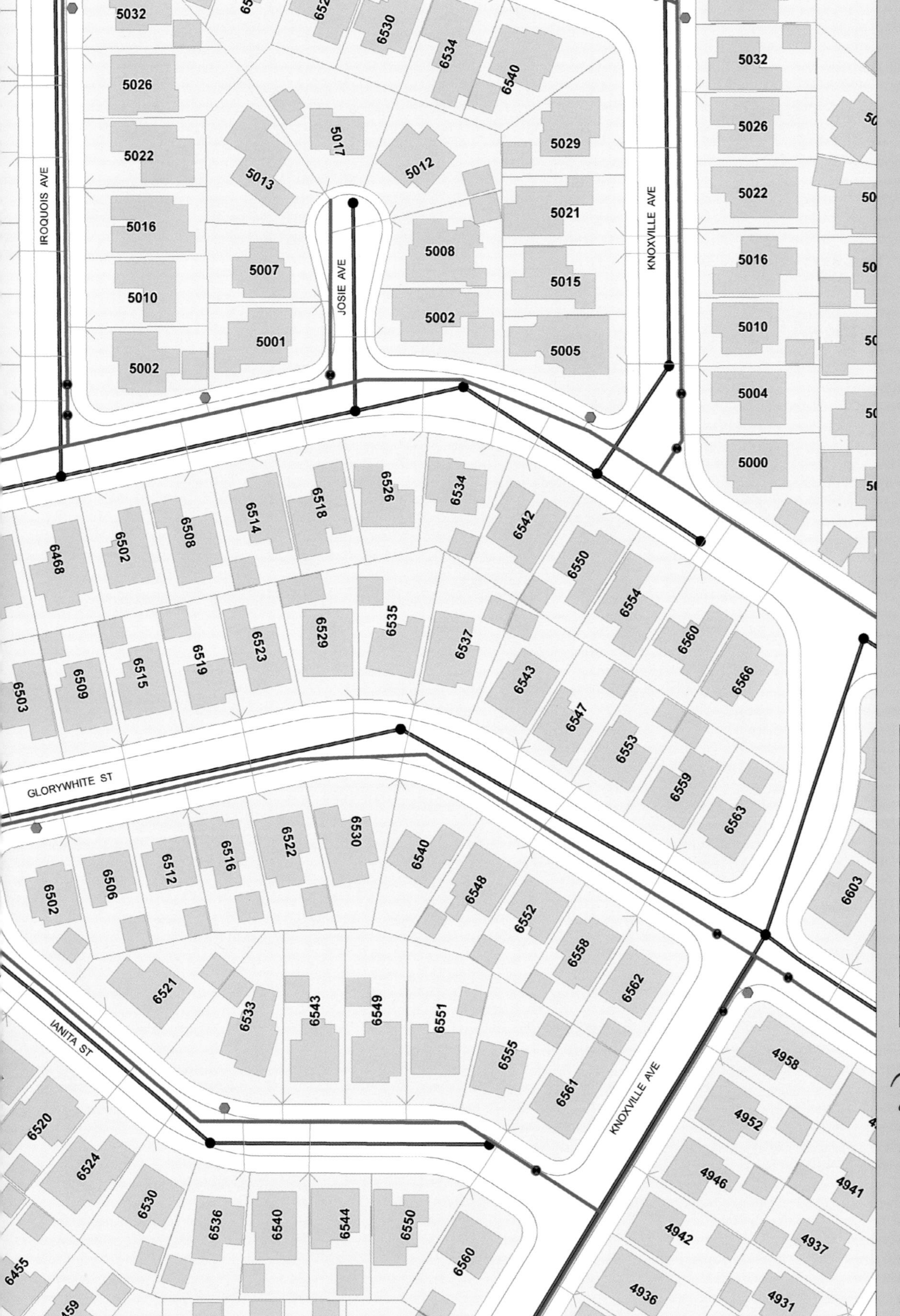

This map demonstrates some of the infrastructure layers developed and maintained by Lakewood's GIS. Some of the layers such as water lines, sewer lines, manholes, and curbs were digitized using a variety of sources that included as-builds and substructure maps. Individual feature layers such as fire hydrants and water valves were captured in the field by water department staff using GPS equipment. Building footprints, transmission towers, and transmission lines were digitized using four-inch color aerials and parcel data. Annotation layers such as addresses and street names were created using the address point and street centerline layers.

Courtesy of the City of Lakewood, California.

Renewable Energy Siting: Collocating Wind Energy and Ethanol Production in Kansas

Tetra Tech EC, Inc.
Bothell, Washington, USA
By Chris Spagnuolo, Ben Fairbanks, Peter Omdal, and Ellen Jackowski

Contact
Andy Bury
andy.bury@tteci.com

Software
ArcInfo 9.1

Printer
HP Designjet 5500

Data Source(s)
Kansas Geospatial Community Commons, National Renewable Energy Laboratory, Pennwell, ESRI

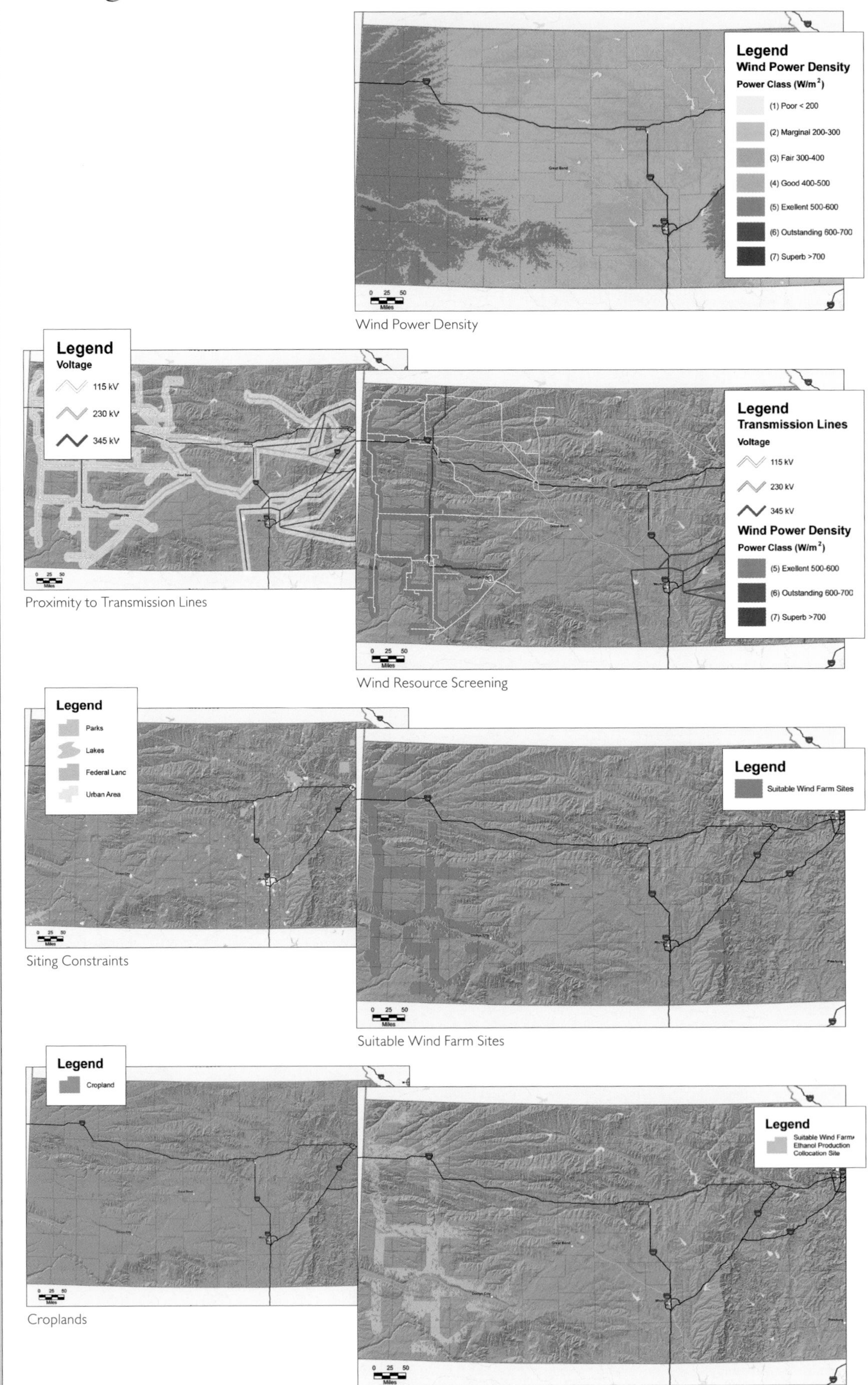

Wind Power Density

Proximity to Transmission Lines

Wind Resource Screening

Siting Constraints

Suitable Wind Farm Sites

Croplands

Collection of Croplands and Wind Farms

This analysis uses GIS to select areas that are most suitable for the collocation of wind turbines and ethanol-producing corn crops. Wind Resource Area (WRA) identification is the first step in identifying these prime collocation sites. Many aspects must be evaluated when identifying and selecting WRAs, including the wind energy resource, proximity to transmission lines, land use, and land management. The second step in identifying collocation sites is to determine the current land-use areas designated as agricultural. The intersection of Wind Resource Areas and agricultural lands identifies the most suitable sites for the collocation of wind turbines and corn crops.

Courtesy of Tetra Tech EC, Inc.

Patrick Engineering, Inc.
Chicago, Illinois, USA
By Brian Fee, James Alberts

Contact
Brian Fee
bfee@patrickengineering.com

Software
ArcGIS Desktop 9.1

Printer
HP Designjet 1050c

Data Source(s)
Village of Winfield

Overall Segment Ratings

Utility Atlas

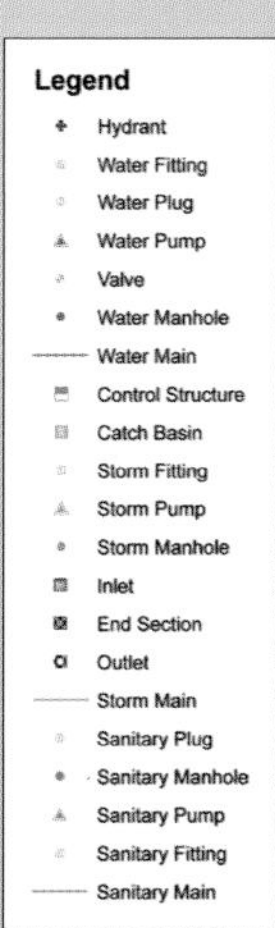

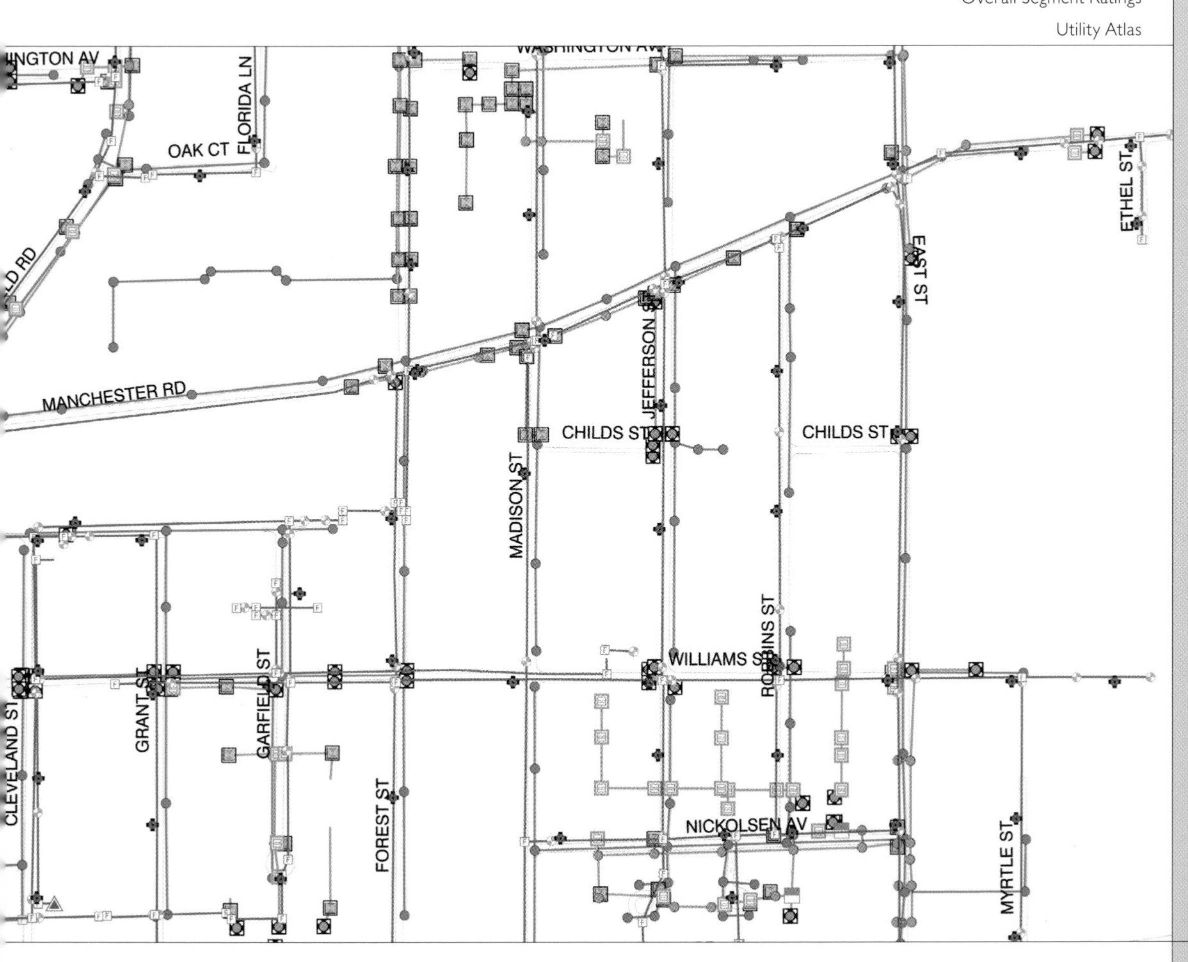

The Village of Winfield, located in Winfield, Illinois, a suburb of Chicago, wanted a systematic method for analyzing where funds would best be spent on capital improvements and maintenance of its roadway and utility infrastructure. Patrick Engineering digitized data from existing plan sets and—working with the village engineers and public works personnel—categorized the condition of roadway surfaces and utility piping. Within the GIS, each feature was located inside a roadway segment. Then the system and overall condition ratings were calculated for each segment. These ratings and utilities were mapped allowing GIS users to use custom reports to determine the replacement cost of any features.

Courtesy of Patrick Engineering, Inc.

Kirkland's Utility Map Books

City of Kirkland
Kirkland, Washington, USA
By Chris Mast, Karl Johansen, Joe Plattner, Kim Sun, Seppo Tervo, and field team

Contact
Xiaoning Jiang
xjiang@ci.kirkland.wa.us

Software
ArcGIS Desktop 9.1

Printer
HP Designjet 5000ps

Data Source(s)
City of Kirkland

2006 Surface Water Map Book

2006 Kirkland Wastewater Map Book

2006 City of Kirkland Water Map Book

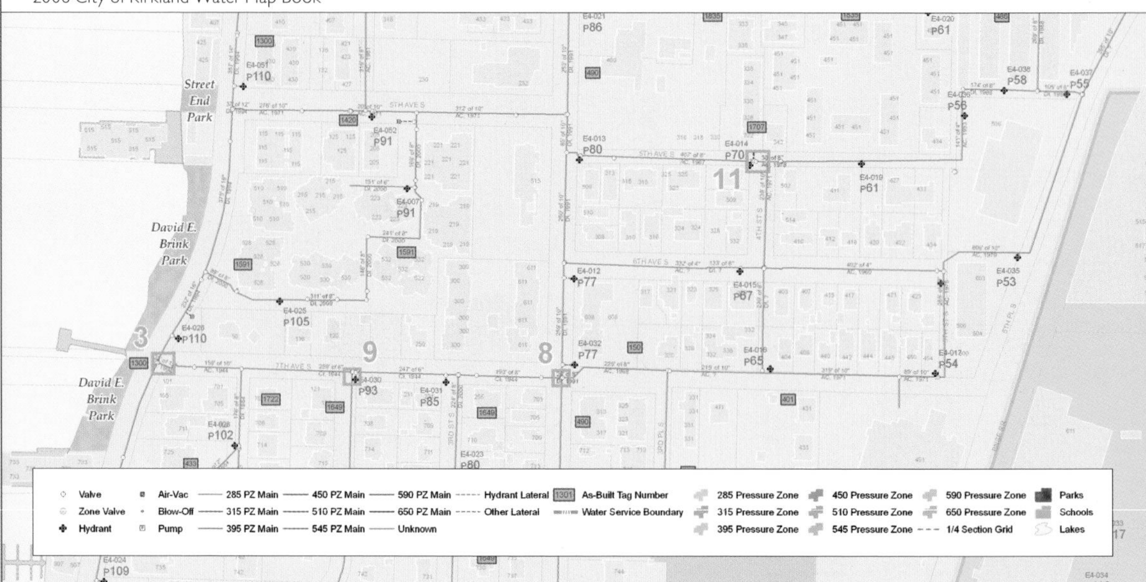

The City of Kirkland, Washington, recently completed the first phase of an ambitious five-year GIS project that included major initiatives in organization, data development, and application integration. Kirkland's GIS staff is currently shifting the program's focus to enhance data quality and accessibility, as well as to promote value-added geographic analysis and displays. Utility infrastructure mapping was one of the top priorities during the first phase. This project identified municipal business functions that would benefit greatly from a GIS inventory of utility infrastructure data. These include routing of field staff, tracking customer information, identifying high-priority maintenance areas, facility inventory, data quality, network tracing, capital project planning, emergency preparedness, and regulatory compliance. The project began with the surface water drainage ("storm") utility, followed by the sanitary sewer system, and finally the water utility.

Benefits of the project included improved data quality, greatly improved data transfer to other business systems, data distribution to outside requestors, and a set of highly successful utility atlases. This positive project outcome serves as a strong foundation for current municipal initiatives such as mobile technology enhancements, GIS–business application integration, and regional government cooperation.

Courtesy of the City of Kirkland, Washington.

Analysis of Constraints to New Jersey Public Drinking-Water Well Locations

New Jersey Department of Environmental Protection (NJDEP)
Trenton, New Jersey, USA
By Edward Apalinski

Contact
Edward Apalinski
edward.apalinski@dep.state.nj.us

Software
ArcGIS Desktop 9.1

Printer
HP Designjet 1055cm

Data Source(s)
NJDEP Water Supply

Groundwater Drinking Water Source Constraint Sum Values (Higher Sum Values = More Constraints)

High : 246

Low : 0

This one-acre grid analysis assesses the New Jersey Department of Environmental Protection's water quality and water quantity constraints in locating new public community drinking-water wells. The groundwater quality constraints, identified from source-water protection strategies, potentially could contaminate source water, requiring a need for increased monitoring and treatment, as well as removal of affected wells from service. The water quantity portion of the analysis maps areas where water-resource-related concerns exist, such as stressed aquifers or endangered species habitat. Each analysis can be used independently for planning future groundwater diversions. However, both analyses cumulatively provide a holistic tool for evaluating water quality and quantity constraints in locating future drinking-water wells.

Courtesy of the New Jersey Department of Environmental Protection, Division of Water Supply.

Index by Organization